Quantum Dots

Synthesis, characterization, and optical investigations

Online at: https://doi.org/10.1088/978-0-7503-5704-3

IOP Series in Coherent Sources, Quantum Fundamentals, and Applications

About the Editor

F J Duarte is a laser physicist based in Western New York, USA. His career has covered three continents while contributing within the academic, industrial, and defense sectors. Duarte is editor/author of 15 laser optics books and sole author of three books: *Tunable Laser Optics, Quantum Optics for Engineers*, and *Fundamentals of Quantum Entanglement*. Duarte has made original contributions in the fields of coherent imaging, directed energy, high-power tunable lasers, laser metrology, liquid and solid-state organic gain media, narrow-linewidth tunable laser oscillators, organic semiconductor coherent emission, N-slit quantum interferometry, polarization rotation, quantum entanglement, and space-to-space secure interferometric communications. He is also the author of the generalized multiple-prism grating dispersion theory and pioneered the use of Dirac's quantum notation in N-slit interferometry and classical optics. His contributions have found applications in? numerous fields, including astronomical instrumentation, dispersive optics, femtosecond laser microscopy, geodesics, gravitational lensing, heat transfer, laser isotope separation, laser medicine, laser pulse compression, laser spectroscopy, mathematical transforms, nonlinear optics, polarization optics, and tunable diode-laser design. Duarte was elected Fellow of the Australian Institute of Physics in 1987 and Fellow of the Optical Society of America in 1993. He has received various recognitions, including the *Paul F Foreman Engineering Excellence Award* and the *David Richardson Medal* from the Optical Society.

Coherent Sources, Quantum Fundamentals, and Applications

Since its discovery the laser has found innumerable applications from astronomy to zoology. Subsequently, we have also become familiar with additional sources of coherent radiation such as the free electron laser, optical parametric oscillators, and coherent interferometric emitters. The aim of this book Series in Coherent Sources, Quantum Fundamentals, and Applications is to explore and explain the physics and?technology of widely applied sources of coherent radiation and to match them with utilitarian and cutting-edge scientific applications. Coherent sources of interest are those that offer advantages in particular emission characteristics areas such as broad tunability, high spectral coherence, high energy, or high power. An additional area of inclusion are the coherent sources capable of high performance in the miniaturized realm. Understanding of quantum fundamentals can lead to new and better coherent sources and unimagined scientific and technological applications. Application areas of interest include the industrial, commercial, and medical sectors. Also, particular attention is given to scientific applications with a bright future such as coherent spectroscopy, astronomy, biophotonics, space communications, space interferometry, quantum entanglement, and quantum interference.

Publishing benefits
Authors are encouraged to take advantage of the features made possible by electronic publication to enhance the reader experience through the use of color, animation and video, and incorporating supplementary files in their work.

Do you have an idea of a book that you'd like to explore?
For further information and details of submitting book proposals, see iopscience.org/books or contact Ashley Gasque at ashley.gasque@iop.org.

A full list of titles published in this series can be found here: https://iopscience.iop.org/bookListInfo/series-in-coherent-sources-and-applications.

Quantum Dots

Synthesis, characterization, and optical investigations

Yarub Al-Douri
Piri Reis University, Istanbul, Turkey & University of Malaya, Kuala Lumpur, Malaysia

IOP Publishing, Bristol, UK

ISBN 978-0-7503-5704-3 (ebook)
ISBN 978-0-7503-5702-9 (print)
ISBN 978-0-7503-5705-0 (myPrint)
ISBN 978-0-7503-5703-6 (mobi)

DOI 10.1088/978-0-7503-5704-3

Version: 20240901

IOP ebooks

British Library Cataloguing-in-Publication Data: A catalogue record for this book is available from the British Library.

Published by IOP Publishing, wholly owned by The Institute of Physics, London

IOP Publishing, No.2 The Distillery, Glassfields, Avon Street, Bristol, BS2 0GR, UK

US Office: IOP Publishing, Inc., 190 North Independence Mall West, Suite 601, Philadelphia, PA 19106, USA

I dedicate this book to my family, parents, brothers, sister and Iraq

Contents

Preface

Scientific research of electronic devices is limited for naturally occurring isolated atoms, particles, metallic, semiconductors, and beta radiation beams. Their engieeernig has three-dimensional systems, while effective geometrical reduction to two dimensions via spatial localization towards plane, line, or point (i.e., confinement of electron in one dimension at de Broglie wavelength), WHICH happens when atoms or electrons localize on imperfect crystals, which means impurities.

In the 1970s, a new era of two-dimensional electronic structure researches of quantum wells (QWs) begun. QWs are very thin and flat, with a higher conduction energy band. The difference between the conduction energy bands of two materials binds electrons in a thin layer. Since the electron's effective mass is small, the de Broglie wavelength is relatively large. The electron's motion bound in several monolayers is two-dimensional, and the perpendicular direction's excitation is strongly quantized. The unique properties of quasi-two-dimensional systems promises optoelectronic applications and have attracted a great deal of research attention, reflecting the rapid development of extensive research and productive technology, leading to integer quantum Hall effect discovery by Klaus von Klitzing's research team who was awarded the Nobel Prize in Physics in 1985. The fractional quantum Hall effect (FQHE) brought the Buckley Prizes to its discoverers: D C Tsui, H C Stormer, and A C Gossard, and to Robert Laughlin for his theoretical works. Meanwhile, the quasi-two-dimensional systems' properties have been investigated and QWs have been implemented for years in numerous devices, such as laser diodes used in CDs players or microwave receivers used in satellite TVs.

In the 1980s, the rapid development of technologies in lithographic techniques has made it possible to confine electrons in quasi-one-dimensional QWs that are synthesized in the form of miniature strips etched in a sample containing a QW. The ability of lithography means that the transverse dimensions are larger than QWs' depth, reaching 10–500 nm. Complete quantization of electron's free motion is implemented by trapping it in a quasi-zero-dimension quantum dot (QD), which was achieved by Texas Instruments Incorporated. Later on, square QDs of 250 nm side length were etched by lithography. Subsequent efforts have reported QD creation in AT&T Bell Laboratories and Bell Communication Research Incorporated, including QDs of less than 30–45 nm diameters.

Due to strong confinement in all three dimensions, QDs systems are similar to atoms, and therefore are frequently referred to as artificial atoms, super-atoms, or QD atoms. What makes QDs unusual objects is, first, the possibility of controlling their shape, dimensions, energy levels structure, and confined electrons number. So, it is possible to investigate and create such models as rectangular or parabolic potential well binding one or several particles (with the same or opposite electric charges), as well as Landau quantization of motion of single electron, the radiative recombination from a few-particle system, and so on. The small number of electrons in QDs facilitates ab initio calculations, which makes them small laboratories of many-body effect and especially attractive for theoretical physicists. Some intriguing effects of two-dimensional electronic systems such as composite fermions have led to FQHE and seem to occur in QDs, where the possibility of numerically solving the Schrödinger's equation may be used to understand them. Current experiments concerned with QDs focus mainly on studying their optical properties (absorption and light emission in visible or far-infrared range, and Raman scattering of light) and electric properties (capacitance and transport studies). Since QDs absorb and emit light in a very narrow spectral range that is controlled by applied magnetic field, they seem to find application in more efficient and precisely controllable semiconductor lasers. The electron energy's strong quantization, with parameters suitable for laser action self-assembled QDs, will allow QD-based lasers to work at higher temperatures and lower injection currents. The possibility of QD application in a new generation of computers is very promising. The small dimensions and dense packing of QDs matrices suggest that they could be used for memory media of huge capacity.

Internationally, researchers are fascinated by semiconductor physics of QDs and their potential application in quantum information technology and photonics. QDs are nanometer-sized three-dimensional structures that confine holes and electrons in dimensions following De Broglie wavelength. The QDs' excitations involve more than single carrier and interaction among carriers, or even dominate the emission properties. So, a simple two-level description is only appropriate under certain well defined experimental conditions. Tremendous progress has been made in understanding their electronic, optical, and spin properties, mainly by performing single dot spectroscopy and using appropriate theoretical models.

The spectacular achievements include controlling quantum coupling in QD molecules, coherent optical manipulation of single electron spins in QDs, controlling cavity reflectivity with a single QD, reaching the strong-coupling regime of interaction between a single QD and a photonic cavity, and the generation of triggered polarization-entangled photon pairs. The exciting applications are robust single and compact, and entangled photon sources with quantum storage devices, ultra-high repetition rates, and basic building blocks such as spin-based quantum information implementations. In addition, high density QDs systems are appealing for classical optoelectronic applications, such as sensitive detectors, modulators, ultra-fast amplifiers, and low-threshold lasers.

Universities, research intititutes, academics, and researchers will benefit from this textbook of 'Quantum dots: Synthesis, characterization and optical investigations' for their undergraduate and postgraduate nano-technology curriculums, as follows: the readers will find an introduction into energetic transitions to describe the crystalline structure and levels of atoms and molecules. In addition to electron interaction (chapter 1), chapter 2 displays the nanoelectronics represented by QDs, giving fundamental concepts, and traditional and new QDs. The characterization, fabrication, preparation, and confinement effect are elaborated under synthesis of QDs as presented in chapter 3. Recent application of QDs as graphene nanostructures with their fabrication methods and their optical properties are explained in chapter 4. Theoretical studies using density functional theory within different approximations, generalized gradient approximation, Engel and Vosko's generalized gradient approximation formalism, and modified Becke–Johnson for elements (Si), compounds (CdS, CdTe PbS, PbSe, and PbTe), and alloys ($CdS_{1-x}Te_x$ and $PbS_{1-x}Te_x$) can be found in chapters 5–9, respectively. Finally, chapter 10 presents a qualitative and new discussion of artificial intelligence-based QDs, including their principles, fundamental concepts, characterisitics, fabrication, preparation, fabrication methods, and optical properties.

Yarub Al-Douri
Piri Reis University, Istanbul, Turkey
University of Malaya, Kuala Lumpur, Malayasia

Author biography

Yarub Al-Douri

Prof. Dr. Yarub Al-Douri is a Fellow of the European Academy of Sciences, Al-Douri is winner of the prestigious Khalifa International Award for Date Palm and Agricultural Innovation 2024. He is one of the Middle-East, North of Africa and Southeast Asia's most renowned scientists known for his contributions in renewable energy and nanotechnology. He is a scientist, researcher, administrator and an excellent lecturer who has students all over the world. Al-Douri has Doctorat D'etat in Materials Science (2000), MSc in Physics (1995) and BSc in Physics (1991). He has been appointed Full-Professor, Visiting Professor, Adjunct Professor, Consultant Expert, Associate Professor, Assistant Professor, Research Fellow (A), Scientific Collaborator and Post-doc in UAE, KSA, Iraq, Malaysia, Turkey, Pakistan, Algeria, Yemen, Singapore, Germany and France, respectively. Al-Douri has initiated Nanotechnology Engineering MSc Program and Nano Computing Laboratory, the first in Malaysia and Southeast Asia, in addition to founding Applied Materials Laboratory in Algeria. Al-Douri was a Head of Department of Nanomaterials and Department of International Networking and Collaboration, additionally to Secretary of Department of Physics. He has received 76 national and international prizes and awards including winner of IAAM Scientist Award by International Association of Advanced Materials, Sweden 2022, winner of World's Top 2% Scientist Career-Long Citation Impact by Stanford University, USA 2020, World's Top 2% Scientists by Stanford University, USA 2023, 2022, 2021 & 2020, OeAD Award, Austria 2020, Japan Society for the Promotion of Science (JSPS) Award 2019, Asian Universities Alliance (AUA) Award 2019, Iraqi Forum for Intellectuals and Academics Award 2019, Best Researcher Award at Cihan University Sulaimaniya 2021 & 2019, Best Paper at Global Conference on Energy and Sustainable Development, Coventry, UK 2015, Distinguished Researcher Award at University Malaysia Perlis 2011-2015, Gold Award at ITEX Kuala Lumpur 2013, TWAS-UNESCO Associateship 2009–2012 & 2012–2015 and others. He has more than 30 years' experience of research, teaching, administrative and editorial board management, organizing events, research grants and consultations, in addition to more than 886 publications with renowned international publishers like Elsevier, Springer, AIP, IOP, Taylor & Francis and Wiley, and US$ 5.1M research grants. Al-Douri has notable citations > 13.2K and h-index > 63. He has supervised more than 24 students. Al-Douri is Associate Editor of Nano-Micro Letters (Springer, Q1, IF = 31.6), Editor-in-Chief of Experimental and Theoretical Nanotechnology (Scopus-indexed), Editor-in-Chief of World Journal of Nano Science and Engineering, Editor and Peer-reviewer of 34 international journals, member of 13 international scientific bodies, organized 33 international conferences, chaired 11 international conferences, appointed 38 internal and external examiner for post-graduates students, presented 41 training courses and public lectures and has been invited 64 times as keynote, invited and guest speakers to many reputed universities in Europe, Asia and Africa that ended up in generating more than 9 MOUs and MOAs. His research field focuses on renewable energy, nanotechnology, nanoelectronics, nanomaterials, modelling and simulation, semiconductors and optical studies. Al-Douri has contributed to academic and research life for upgrading universities to the international level. Finally, Al-Douri is a public figure at international media in UK, Singapore, Malaysia, Qatar and UAE.

Chapter 1

Energy states

Energy states refer to the distinct levels of energy that a physical system, such as an atom, molecule, or nucleus, can occupy. These states are quantized, meaning that the system can only possess specific, discrete energy values rather than a continuous range. The concept of energy states is fundamental in quantum mechanics and is used to explain phenomena such as atomic spectra, molecular vibrations, and nuclear reactions.

So, energy states are associated with electron configurations around the nucleus. Electrons can move between energy levels by absorbing or emitting photons, which correspond to specific wavelengths of light. This process results in characteristic emission or absorption spectra for each element.

1.1 Energetic transition

A quantum mechanical system or particle that is bound can only take on certain discrete values of energy. The energy levels have discrete values that are utilized commonly for molecules, ions and electrons constitute the energy levels, and meantime are bounded by the nucleus electric field and refer to nuclei energy levels or rotational or vibrational energy levels at molecules. Here arises a quantized term that means the spectrum of the energy system with levels of discrete energy. The negative potential energy for usual convention and bound electron states comes from zero potential energy at unlimited distance from the molecule or atomic nucleus.

The ground state term means that the molecule, ion, or atom is in the lowest level of energy. While the excited state means that the molecule, ion, or atom has a higher level of energy, which is the electron has a higher energy similar at the ground state. The degenerate state has many quantum mechanical states in the same energy, which are called degenerate energy levels.

doi:10.1088/978-0-7503-5704-3ch1

1.2 Atoms

1.2.1 Intrinsic levels of energy

For the electrons' equations of energy at different levels in the atom, the zero energy is fixed when the electron has left the atom, i.e., when the principal quantum number of the electron has infinite value ($n = \infty$). The electrons are considered lower in energy and negative when electrons are bounded to others at a closer value of principal quantum number (n).

Assume there is one electron in a given atomic orbital in a hydrogen-like atom (ion). The energy of its state is mainly determined by the electrostatic interaction of the (negative) electron with the (positive) nucleus. The energy levels of an electron around a nucleus are given by:

$$E_n = -hcR_\infty \frac{Z^2}{n^2} \tag{1.1}$$

(typically between 1 and 103 eV), where R_∞ is the Rydberg constant, Z is the atomic number, n is the principal quantum number, h is Planck's constant, and c is the speed of light. For hydrogen-like atoms (ions) only, the Rydberg levels depend only on the principal quantum number n.

This equation is obtained by combining Rydberg's formula for any hydrogen-like element (shown below) with $E = h\nu = hc/\lambda$, supposing that the principal quantum number of electron (n) is $n1$ in Rydberg's formula and $n2 = \infty$ (principal quantum number of electron coming down within emitted photon). Rydberg's formula is deduced from the emission data of the empirical spectroscopic.

$$\frac{1}{\lambda} = RZ^2\left(\frac{1}{n_1^2} - \frac{1}{n_2^2}\right) \tag{1.2}$$

The time-independent Schrödinger's equation can help to provide equivalent formula with a Hamiltonian operator of kinetic energy utilizing the wave function as an eigenfunction for obtaining eigenvalues of energy levels, but other fundamental physics constants could replace Rydberg's constant.

The electrostatic interactions of electrons have multi-electron atoms. The interactions of electron–electron around the atom could increase the energy level, which are neglected if the spatial electronic overlapping of the wave function is low. The mentioned equation (1.2) is justified by electronic interactions of multi-electron atoms to be inaccurate, as indicated with the atomic number (Z). The shielding effect is an interpretation of this phenomenon, where the inner electrons are bounded to the nucleus neglecting its charge and outer electrons represent the reduced charge for an effective nucleus. This leads to an approximated correction where atomic number (Z) is substituted by molecule affect (Z_{eff}), i.e., effective nuclear charge depending on principal quantum number of electron (n).

$$E_{n,l} = -hcR_\infty \frac{{Z_{\text{eff}}}^2}{n^2} \tag{1.3}$$

The azimuthal quantum number of electron (l) and their levels within Z_{eff} affect different atomic electron of energy levels [1]. The Aufbau principle guides to fill-in atom by electrons to achieve the electron configuration, which takes various energy levels into consideration.

1.2.2 Splitting of fine structure

The Darwin term (interaction of *s* shell electrons inside the nucleus), spin–orbit coupling (electrodynamic interaction between nucleus's electric field and electron's spin and motion) and relativistic correction of kinetic energy are an impetus to arise fine structures, which affect the levels of magnitude; 10^{-3} eV.

1.3 Molecules

The atomic situation becomes more stable depending on the chemical bonds between the atoms in any molecule, which means that the total levels of energy for the atoms in a molecule are less than if the atom is not bonded. For the separate case, each one has a covalent bond and its orbital affects each other energy level to create the bonding and antibonding orbitals of the molecule. It is well-known that the energy level of antibonding is high and energy level of bonding orbital is low. For the molecular bonds to be stable, the covalent bond of the electron occupies a lower bonding orbital, which could be indicated by symbols π or σ depending on its situation. But for an antibonding orbital, it could be indicated by π^* or σ^* orbitals. There is another case of a non-bonding orbital, which has an orbital with an electron in an outer shell that does not contribute to the bonding and the level of energy of the constituent atom. These orbitals are *n* orbitals. The electrons in an *n* orbital are lone pairs [1]. For the polyatomic molecule, different rotational and vibrational levels of energy are involved.

For the molecular state of energy, the eigenstate of the molecular Hamiltonian is a summation of translational, nuclear, rotational, vibrational, and electronic components, such that:

$$E = E_{\text{electronic}} + E_{\text{vibrational}} + E_{\text{rotational}} + E_{\text{nuclear}} + E_{\text{translational}}$$

where the eigenvalue of the electronic molecular Hamiltonian refers to $E_{\text{electronic}}$ (a value of potential surface of energy) at molecular equilibrium geometry. The energy level of the molecule is labeled by molecular symbols.

The specific energy of the components varies with the specific state of energy and its substance.

In quantum chemistry and molecular physics, the level of energy is quantized of a bound state of quantum mechanics. There are different types of diagrams for the energy levels for atomic bonds in the molecule.

1.3.1 Level of energy

The molecular electron and atomic electron transitions present an increasing level of energy from E_1 to E_2, which results in the absorption of a photon, indicated by a red squiggly arrow, and its energy is decreasing in level from E_1 to E_2, which results in

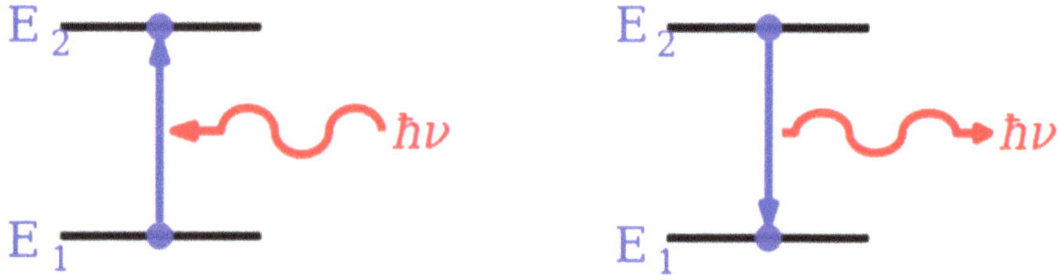

Figure 1.1. Transition of energy between atomic levels.

the emission of a photon, indicated by a red squiggly arrow whose energy is $h\nu$ (figure 1.1).

The electrons in the molecules and atoms can vary their levels of energy by absorbing or emitting of photons (electromagnetic radiation), where their energies equal the difference of energy between two levels. These electrons could be removed from chemical species such as ions, molecules, or atoms. The complete process of removal of electrons from any atom could be in an ionization form, which moves the electrons out to another orbital, including infinite principal quantum number of electrons, so far away so as to have no effect on other remaining atoms (ions). There are 1st, 2nd, and 3rd levels for different atoms. The ionization energy of removal is the 1st, then 2nd, then 3rd level of highest electronic energy, respectively, of the atom in the ground state. The energy corresponds to an opposite quantity that can be released in photon energy form when the electrons are added to positively charged atoms or ions. The molecules undergo a transition in their rotational or vibrational levels of energy. The energy transition of a level can be nonradiative, which means that the absorption or emission of photons are not involved.

If the molecule, ion, or atom has a lowest level of energy, its electrons are in the ground state. If it has a highest level of energy, it is an excited state. Such a case can be excited to a highest level of energy by absorbing photons equal to the difference of energy between levels. In contrast, the excited states could go down to lower levels of energy by emitting photons equal to the difference of energy. The energy of the photons equals Planck's constant (h) multiplied by its frequency (ν), which is correlated directly to its frequency and inversely to its wavelength (λ) [2].

$\Delta E = h\nu = hc/\lambda$, where c is the speed of light. In addition, different types of spectroscopy are based on detecting wavelength or frequency of absorbed or emitted of photons to providing information about a material's electronic levels of energy by analyzing the spectrum.

It is known now that the asterisk is indicated for the excited state. So, the electronic transition of molecular bond from ground to excited states is referred to as $n \rightarrow \pi^*$, $\pi \rightarrow \pi^*$ or $\sigma \rightarrow \sigma^*$, which means an electronic excitation from σ bonding to σ antibonding orbitals, from π bonding to π antibonding orbitals, or from n non-bonding to π antibonding orbitals [1, 2]. Contrary electronic transition for the mentioned types of molecules is possible to come back to their ground states, which is referred to as $\pi^* \rightarrow n$, $\pi^* \rightarrow \pi$ or $\sigma^* \rightarrow \sigma$.

The electronic transition level of the energy of a molecule could be combined with the vibrational transition, which is named vibronic transition. The rotational and vibrational transitions could be combined via rovibrational coupling. In rovibronic coupling, the electronic transition is combined with both rotational and vibrational transitions. The photons are involved in transitions that may have different ranges of energy in an electromagnetic spectrum, such as microwave radiation, infrared, visible light, ultraviolet, and x-ray, depending on a transition type. Generally, the energy difference of level the electronic states is large, between vibrational levels is intermediate, and between rotational levels is small despite of overlapping. The translational levels of energy are continuous and could be regarded a kinetic energy utilizing classical mechanics. While the highest temperatures cause fluid molecules and atoms to move quickly, increasing their translational energy, and thermally to excite the molecules for a highest average of amplitude of rotational and vibrational modes (exciting molecules to a highest level of energy). This means that as temperature increases, the rotational, vibrational, and translational contributions to a molecular capacity of heat let the molecule absorb heat and hold more energy. The conduction of heat occurs when the atoms or molecules transfer the heat between each other. While at the highest temperature, the electron can be excited to highest orbital of energy in molecules or atoms. The dropping of electrons to a lower level of energy can emit energy as a photon, releasing a colored glow. The electrons far from the nucleus have a higher potential energy than electrons near to the nucleus, and therefore they become lighter bounded to the nucleus, which means that their potential energy is negative and inversely dependent on their distance from the nucleus [3].

1.3.2 Crystalline materials

Solids have crystalline forms that have energy bands, which are replaced by energy levels. The band, any band, can take any electron of a specific energy. First, it is appeared to be exceptional for the energy level's requirements. Anyway, band theory shows that energy bands have different discrete levels of energy that are very near to each other for resolving. For one band, there are numbers of levels, like the numbers of atoms in a crystal, so, although these electrons are restricted to their energies, they like to take continuous values. To clarify, the more important levels of energy in the crystal are located on the top of the valence band, while the conduction band is located in the bottom; on the other hand, the energy levels, vacuum level, and Fermi level are of any states of defect in the crystal.

In this chapter, the conditions of any nanocrystal that can be counted as quantum dots (QDs) are discussed. To elaborate the quantum confinement of three-dimensions, the problem of quantum mechanics of the particle motion in any box will be recalled. This is a simple model that has been ameliorated during the previous years by adding further and difficult expressions to Hamiltonian, taking into consideration the Coulomb interactions of electron hole, non-parabolic bands, or complicated structure of the valence band. This is followed by a development of different models that are the result of different theories for comparison with the experimental data.

Linear absorption is not enough to unambiguously identify the states of energy for QDs. The combination of nonlinear and linear optical methods permits an immediate comparison of experiment and theory. Starting with states of one-electron-hole-pair (lEHP) and continuing for two-pair (2EHP), followed by many-particle states, the population and density of excited states in the QDs are increasing. This could be understood by the increasing density of excitation or changes in the nanocrystal size. So, it is inevitable to discuss a weak, intermediate, and strong regime of confinement.

1.4 States of one-electron–hole-pair

1.4.1 The particle-in-the-box model

Nanocrystals with spatial extension are considered larger than the lattice constant. This is very similar in the sizes range; the bulk materials have a crystalline structure that has been improved and a quantum confinement effect can be noticed. The quantum confinement description within the effective mass approximation (EMA) framework is an accepted approach. Alternatively, the linear combination of atomic orbitals approximation is approached. The states of energy computed via tight-binding approximation are fitted well with other results via EMA in the range of QDs radii going down to 1.5 nm, as illustrated [4–6]. Wang and Zunger [7] have results beyond EMA in the small sizes of QDs. They have achieved an agreement between experimental data and pseudopotential calculations results for CdSe QDs of sizes for a small cluster ($R = 0.6$ nm) up to CdSe QDs of 1000 atoms ($R = 1.9$ nm).

For the bulk scale, the states of electrons are obtained by solving Schrodinger's equation of stationary for electrons in a periodic potential,

$$\hat{H}\psi(r) = \left[-\frac{\hbar^2}{2m}\nabla^2 + V(r)\right]\psi(r) = E\psi(r) \tag{1.4}$$

The underlined periodicity of Bravais lattice refers to the potential $V(r)$:

$$V(r) = V(r + \boldsymbol{R}) \tag{1.5}$$

Related to all lattice vectors $\boldsymbol{R}$. The eigenstates Yi of (3.1) are formed of enveloped functions, function $U_{v,k}(r)$ and plane wave e^{ikr} periodically with Bravais lattice vector

$$\psi_{v,k}(r) = e^{ikr}\, u_{v,k}(r) \tag{1.6}$$

and

$$u_{v,k}(r) = u_{v,k}(r + R) \tag{1.7}$$

The band index refers to v, and the reduced wave vector of the first Brillouin zone refers to k. The Bloch theorem is expressed in equations (1.6) and (1.7). The parabolic band approximation of energy eigenvalues is given;

$$E(k) = \frac{\hbar^2 k^2}{2m} \tag{1.8}$$

where electron or hole effective mass refers to m.

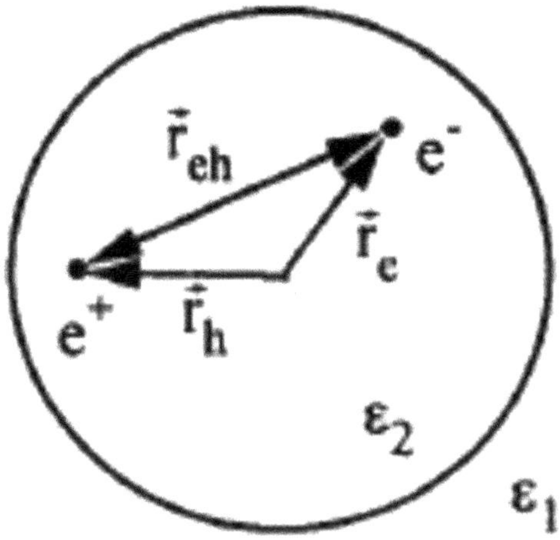

Figure 1.2. Hole and electron in a sphere of semiconductor with dielectric constant (ε_2) embedded in host material with (ε_1). Reprinted from [4], Copyright (2022), with permission from Elsevier.

Figure 1.2 shows the electron-hole-pair state problem inside a spherical potential. Simply, the sphere of semiconductor of radius R is surrounded by infinite barriers of high potential demonstrated by matrix material. In a framework of approximation of the envelope function, it is supposed that the wave function expands in a result of parts of periodic cell $U_{v,k}$ of Bloch functions with a specific and new envelope function. The Bloch function periodic parts are assumed to be in the same well and barrier,

$$u_{v,k}(r)\text{barrier} = u_{v,k}(r)\text{well} = u_{v,k}(r) \tag{1.9}$$

The goal for determining the particle-in-a-spherical-box problem is a new envelope function Ψ for holes and electron,

$$\psi(r) = \psi(r)u(r) \tag{1.10}$$

The Hamiltonian operator of envelope function $\Psi(r)$ in a single approximation of parabolic band, neglecting the first Coulomb interaction, is:

$$\hat{H} = -\frac{\hbar^2}{2m_e}\nabla_e^2 - \frac{\hbar^2}{2m_h}\nabla_h^2 + V_e(r_e) + V_h(r_h) \tag{1.11}$$

with a confinement potential

$$V_i(r_i) = 0 \text{ for } r_i\langle R(i = e, h) \text{ or } \infty \text{ for } r_i\rangle R(i = e, h) \tag{1.12}$$

For non-interacting electron-hole pairs, the envelope wave function Ψ can be expressed as separate contributions from the electrons and holes,

$$\psi(r_e, r_h) = \varphi_e(r_e).\ \varphi_h(r_h) \tag{1.13}$$

The solution to Schrödinger's equation for (1.11–1.13) can be found in [8]. The normalized wave functions Φ_i for holes and electrons are:

$$\varphi^{i}_{nlm}(r) = Y_{lm}\sqrt{\frac{2}{R^3}\frac{J_l\left(\chi_{nl}\frac{r}{R}\right)}{J_{l+1}(\chi_{nl})}} \tag{1.14}$$

with $-l \leqslant m \leqslant l$; $l = 0, 1, 2,\ldots$; $n = 1, 2, 3, \ldots$. Thus, the Bessel functions are denoted by J_l, and the spherical harmonics are denoted by Y_{lm}. The energy eigenvalues E_{nl} correspond to the wave function, which must vanish at $r = R$, the boundary of the QDs,

$$J_l\left(\chi_{nl}\frac{r}{R}\right)\Big|_{R=r} = 0 \tag{1.15}$$

In addition, the condition is determined:

$$E^{e,\,h}_{nl} = \frac{\hbar^2}{2m}\frac{\chi^2_{nl}}{R^2} \tag{1.16}$$

X_{nl} represents the nth zero of the spherical Bessel function of order l, and $m_{e,h}$ denotes the effective masses of the electron or hole. The nanocrystal's radius is denoted by R. The quantum numbers are labeled as $l = 0, 1, 2, \ldots$ with corresponding letters $s, p, d, \ldots$. The first roots are $X_{1s} = \pi$; $X_{1p} = 4.493$; $X_{1d} = 5.763$; $X_{2s} = 2\pi$; $X_{2p} = 7.725$, etc. For equation (1.16), the quantization energy of the lowest energy state, where $n = 1$ and $l = 0$, is given as follows:

$$E^{i}_{10} = \frac{\hbar^2}{2m}\frac{\pi^2_l}{R^2} \tag{1.17}$$

The energy of a particle in a spherical potential takes on discrete values that are inversely proportional to the square of the radius R. This is related to the semiconductor's bandgap, where the increase in the transition energy ΔE is given by:

$$\Delta E = \frac{\hbar^2}{2\mu}\frac{\pi^2}{R^2} \tag{1.18}$$

For the lowest confined lEHP state, the reduced effective mass μ of the electron-hole pair is given by ($\mu = m_{e\,+\,h}/(m_e + m_h)$). This discussion considers an envelope problem. The Bloch wave functions are approximated using a two-band semiconductor model with direct energy gaps, isotropic bands, and parabolic conduction and valence bands with their maxima located at $k = 0$ [8]. However, QDs are often made from semiconductors with indirect bandgaps, such as those with wurtzite or zinc-blende structures, which are grown in highly anisotropic lattice symmetry directions. Therefore, the Bloch wave function $U_{v,k}$ (r) exhibits properties that must be considered to understand the optical behavior of QDs. This will be discussed in the following sections.

1.5 Coulomb interaction

The next step in a thorough analysis involves considering the Coulomb interactions between the hole and electron within the QD. This is described by the following Hamiltonian:

$$\hat{H} = -\frac{\hbar^2\nabla_e^2}{2m_e} - \frac{\hbar^2\nabla_h^2}{2m_h} - \frac{e^2}{\varepsilon\,|r_e - r_h|_2} + V_e(r_e) + V_h(r_h) \tag{1.19}$$

In bulk materials, the Coulomb potential is essential and is responsible for the existence of excitons. The Hamiltonian in equation (1.19), without the confining potentials V_h and V_e, can be separated into the center-of-mass and relative motion of the electron-hole pair. For spherical QDs, the Coulomb potential must be considered, introducing a symmetry breaking due to its dependence on the spatial distance between the electron and hole. Analytical solutions are challenging because the Hamiltonian in equation (1.19) cannot be easily separated into center-of-mass and relative motion components. Thus, the Coulomb energy scale, which is inversely proportional to the electron-hole distance ($\approx 1/R$), should be considered, while the kinetic energy scale is proportional to the square of the inverse radius ($\approx 1/R^2$) [8, 9].

In the strong confinement regime ($R \lll a_B$, where a_B is the excitonic Bohr radius), the QD's small radius allows for the neglect of Coulomb interactions, simplifying the problem to non-interacting electrons and holes. However, for QDs with sizes $R \gg a_B$, the problem can be approximated by a particle-in-a-box model, treating the electron-hole pair (exciton) as a single particle with confined potential acting on its center-of-mass motion. This weak confinement approximation is suitable for semiconductors with high exciton binding energies, such as CuBr and CuCl, which have exciton Bohr radii of 1.25 and 0.7 nm, respectively.

Although these approximations are crude, the Hamiltonian in equation (1.19) has been addressed using various methods, including matrix diagonalization [10], Monte Carlo techniques [11], variational calculations [12], and perturbation theory [13]. When treating the problem using perturbation theory, the lowest excited state energy is given by:

$$E_{10} = \frac{\hbar^2\pi^2}{2R^2}\left[\frac{1}{m_e} + \frac{1}{m_h}\right] - \frac{1.8e^2}{\varepsilon_2 R} \tag{1.20}$$

Transitioning from a non-interacting particle model to an electron-hole pair model alters the description of optical transitions, as illustrated in figure 1.2 compared to figure 1.3.

To address the symmetry breaking issue in the intermediate confinement range (where ($a_{B\,\approx}\,R$), Ekimov *et al* [14] proposed an ansatz by positioning the hole at the center of the QD. They modeled the exciton with a donor-like configuration inside the QD, and variational calculations were employed to define ground state energies as a function of size scale. This localization of the hole suggests that the resulting potential experienced by the electrons is confined within the QD potential.

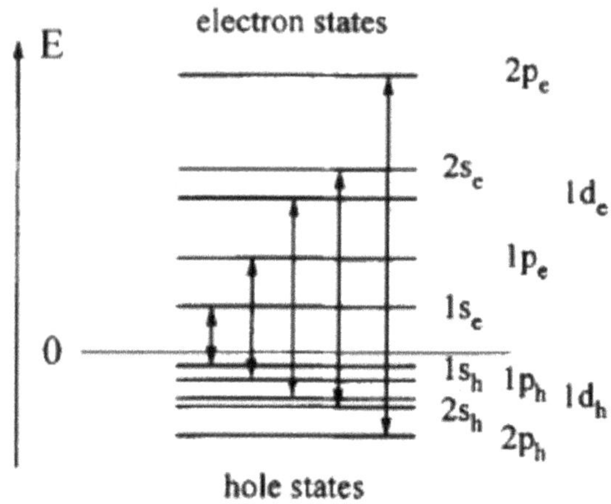

Figure 1.3. Scheme of dipole-allowed optical transitions within non-interacting approximation of particles. Reprinted from [10], with the permission of AIP Publishing.

In numerical methods such as variational calculations, the key innovation lies in the choice of trial functions. For instance, vibrational calculations have been applied to band structures with single parabolic potentials and infinitely high barriers [15]. Trial wave functions incorporate solutions involving non-interacting particles, often utilizing lower-order Bessel functions or exponential functions akin to the 1s-wave function of the hydrogen atom. Typically, these trial functions involve up to three parameters, which are optimized to fit the variational approach. Nair *et al* [15] computed the variational ground state energy of electron-hole pairs using a Hylleraas-type wave function with three parameters. Kayanuma [16], on the other hand, employed polynomial bases for trial wave functions, optimizing the expansion coefficients as variational parameters.

In cases of strong confinement, analytical formulas are valuable for first-principles calculations, as outlined in current research. Assuming the Coulomb potential acts as a small perturbation, one can derive results similar to equation (1.20).

$$E_{10} = \frac{\hbar^2\pi^2}{2R^2}\left[\frac{1}{m_e} + \frac{1}{m_h}\right] - \frac{1.786e^2}{\varepsilon_2 R} - 0.284E^*_{Ryd} \tag{1.21}$$

where E^*_{Ryd} is exciton binding energy in (meV).

In both studies [15, 16], the vibrational values predict a higher energy shift than what is observed in experiments. The main reason for this discrepancy lies in the use of infinitely high potential barriers, which significantly reduce wave functions at the boundaries. A further improvement in the theory involves considering finite barriers in the confined potentials $V_h(r_h)$ and $V_e(r_e)$, as discussed in [17]. Introducing finite barriers helps mitigate the overestimation of the confinement-induced blue shift observed with the ansatz of infinite barriers.

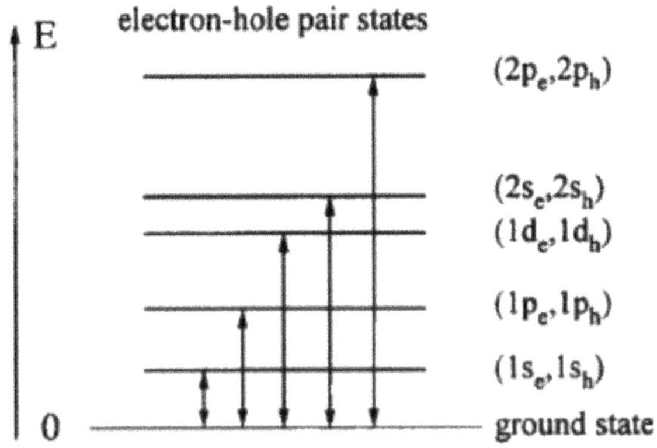

Figure 1.4. Scheme of optical transitions in frame of one-hole-electron pair. Reprinted from [10], with the permission of AIP Publishing.

Figure 1.4 illustrates the results obtained for various values of the finite potential wells V_h and V_e. Additional parameters adopted for CdS QDs include a hole-to-electron mass ratio of $m_h/m_e = 4$, an exciton binding energy of $E.\ R_y = 30$ meV, and energy gaps used for calculations: $E_{\mathrm{glassg}} = 7$ eV, $E_{\mathrm{gCdS}} = 2.58$ eV, with a band offset of 50%/50%. The radius is given in units of the excitonic Bohr radius a_B.

Thoai *et al* [18] considered the difference in effective mass between holes and electrons inside and outside the QD. They assumed $V_{\mathrm{h}} = \infty$, with a finite barrier for electrons, and varied the electron mass outside the QD from $m_e = m_0$ and $m_e = \infty$ depending on the mobility of the matrix material. The confined quantum states were computed as functions of the QD radius.

The computational results were compared with experimental data for CdS QDs in colloidal form, as investigated by Weller *et al* [19], depicted in figure 1.5. For liquid media, it is suggested that electrons exhibit high mobility ($m_e = m_0$). A good agreement was observed for $V_{\mathrm{e}} = 40\ E_{Ry}$, corresponding to a potential height of 1 eV for a liquid solution as the surrounding matrix material.

Wave functions can be obtained by extending numerical methods such as matrix diagonalization techniques, as demonstrated by Hu *et al* [20]. When compared to results from vibrational methods, significant differences in single-pair energy are observed in the range $R/a_B < 1$.

Moreover, changes in radial distribution have been observed in the calculated wave functions of holes and electrons due to the inclusion of Coulomb interactions and consideration of finite barriers. The Coulomb interactions between holes and electrons perturb their wave functions, causing heavier particles to be localized toward the center of the QD.

As holes and electrons penetrate the barriers, their wave functions become nonzero, affecting the probability of dipole-forbidden transitions. The inclusion of Coulomb potentials corrects the confinement-induced energy, leading to shifts towards lower values or modifications in selection rules for transitions.

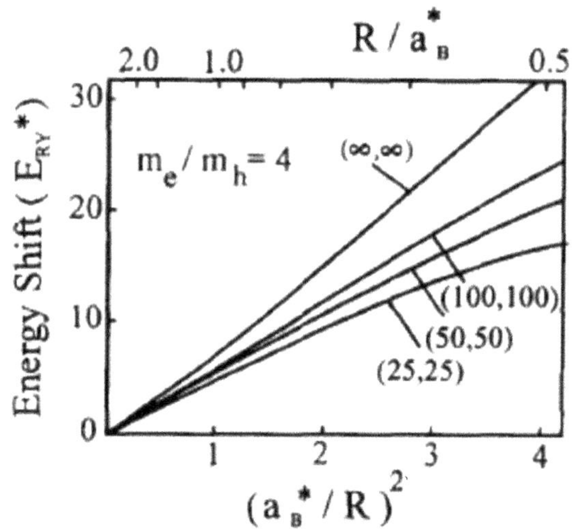

Figure 1.5. Calculated high-energy shifts of hole-electronpair ground state with decreased radius. In brackets, height of potential (V_e, V_h) is referred in Rydberg energy E_{Ry}. Reprinted from [10], with the permission of AIP Publishing.

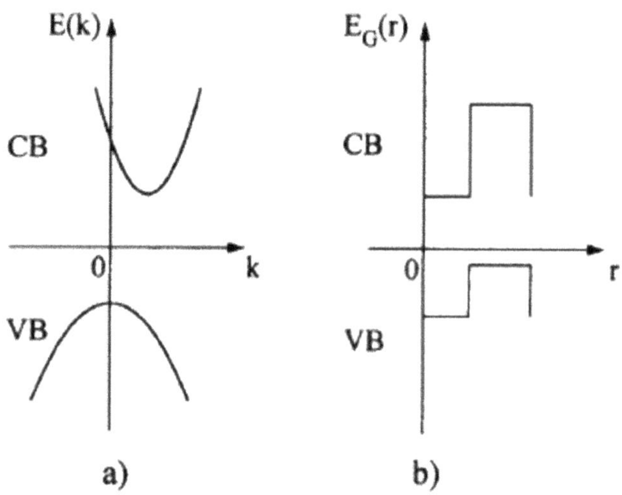

Figure 1.6. Indirect bandgap QDs with respect to (a) momentum space and (b) coordinate space. Reprinted from [10], with the permission of AIP Publishing.

1.5.1 Quantum dots of indirect-type

The electronic properties of semiconductor quantum structures represented by indirect bandgap may emerge from difference in *r*-space and *k*-space between lowest hole and electron wave functions. The varied types of indirect bandgap are illustrated in figure 1.6. The quantum confinement of small artificial structures for indirect bandgap of germanium and silicon has been researched with an arrangement of various semiconductors of dot-like structure of a succession of band offsets similar to type-II quantum wells [21].

The quantum structure manifests their indirect bandgap character by slowing down the recombination rate of optical transition. Therefore, an indirect bandgap quantum structure design aims for engineering probabilities of radiative transition and oscillator strengths. The optical transition of indirect bandgap requires phonons' participation to fulfill the rules of k-selection. For the semiconductor quantum structure of the indirect bandgap, the supposition is a probability of zero-phonon transition that may be enhanced as a result of momentum states mixing both envelope and Bloch parts of the wave functions.

Tolbert *et al* [21] investigated the increase in oscillator strength with decreasing dimensionality in CdSe QDs, which exhibit both indirect and direct bandgaps, unlike group-IV elements like Ge and Si. In bulk CdSe, a solid-solid phase transition occurs under pressure between the rock salt structure (with an indirect bandgap) and the wurtzite structure (with a direct bandgap). At around 3 GPa pressure, CdSe can reversibly transition from wurtzite to rock salt structure, where the optical energy of the rock salt phase is around 0.67 eV in the infrared spectrum. This transition allows for direct changes in optical transitions from direct to indirect bandgaps.

Tolbert and Alivisatos [22] confirmed the presence of structural phase transitions in CdSe nanocrystals with radii between 1 and 1.7 nm using SAXS, x-ray diffraction, and TEM measurements. They also studied the strength of oscillators in optical studies. The probability of intrinsic electronic transitions is influenced by trapping effects, and the absorption spectra were compared and investigated for both structural modifications. In wurtzite QDs, the absorption feature arises from quantum confinement, which is quite distinct. However, this quantum confinement effect diminishes during the structural phase transition under 7 GPa pressure, resulting in a smoother absorption spectrum with a gradual slope at lower energies.

Experimental observations indicate no significant variation in the ratio of indirect to direct bandgaps within the range of QD radii from 0.96 to 1.73 nm. Therefore, while there is a noticeable change in oscillator strength for indirect bandgaps, it does not permanently result in a direct bandgap transition.

Another type of indirect quantum structure involves distinct spatial distributions of electron and hole wave functions, where they reside in different materials. In type-II QDs, for instance, the electron is confined within the dot material while the hole resides in the barrier (matrix) material (see figure 1.6). The Hamiltonian governing this system is essentially similar to equation (1.19), with specific forms for the confinement potentials V_e and V_h.

For type-II behavior to manifest, it is crucial that V_e and V_h exhibit the same functional dependence on the spatial coordinate at the dot boundary, meaning they either both increase or both decrease with r. The energy states, obtained as solutions of the Hamiltonian (1.19) with modified expressions for V_e and V_h, reflect the interaction between the attractive Coulomb force and the separating potential barriers.

Laheld *et al* [23, 24] have calculated the binding energy of hole-electron pair of QDs for infinite barriers of potential and later [25] for problems with finite barriers.

Laheld *et al* [23] have elaborated analytical approximations for limited cases of large and small QD radii to carry out vibrational computations. In the case of very small quantum radii, the problems resemble a textbook problem for finding hole binding energy in the acceptor field to represent the electron in its discrete ground states, although the boundary conditions at $|rh| = R$ for wave functions. For QDs larger than Bohr radius, the result is strongly affected by the ratio of hole and electron masses. The principal result of calculations is a tendency towards smaller hole-electron-pair binding energy in a type-II structure compared with a type-I one structure.

The finite barriers extension by Laheld *et al* [25] highlight an interesting question of probabilities of transition because the overlapping of hole and electron wave functions is very sensitive to the band offset of the material barrier. The particle wave functions are able to penetrate the QD boundary. As size/potential height decreases, the electron wave function accesses the barrier more and overlaps with the hole state, thus increasing the strength of the oscillator. From efforts of Laheld *et al* [25], a quadratic increase of the oscillator strength has been achieved as radius R decreases. While for binding energy, one gets similar results to two-dimension superlattices: binding energy rises as QD radius decreases to a maximum value, followed by a sharp decrease below $R < a_B$. To discuss this phenomenon, the smearing out idea of charges distribution is created by confined particles at very small radii. The maximum value of binding energy corresponds to the minimum extension of the confined particle's wave function.

Problems

1. What are energy states? And, how do they relate to the behavior of subatomic particles?
2. What is the difference between ground state and excited state energy levels? And, how do particles transition between them?
3. How do energy states relate to the emission and absorption of light by atoms and molecules?
4. How do energy states affect the properties of materials, such as their electrical conductivity or magnetic behavior?
5. Can quantum mechanics be used to predict the energy states of large, complex systems, such as molecules or solids?

References

[1] Kittel C 1979 *Introduction to Solid State Physics* 5th edn (New York: Wiley)
[2] Omar M A 1994 *Elementary Solid State Physics: Principles and Applications* 4th edn (New York: Addison-Wesley)
[3] Harison W A 1989 *Electronic Structure and the Properties of Solids* (Toronto: General Publishing Company)
[4] Zhang Q, Mirzaei A, Wang Y, Song G, Wang C, Besteiro L V, Govorov A O, Chaker M and Ma D 2022 *Appl. Catal.* B **317** 121792
[5] Einevoll G T 1992 *Phys. Rev.* B **45** 3410

[6] Ramaniah L M and Nair S V 1993 *Phys. Rev.* B **47** 7132
[7] Wang L-W and Zunger A 1996 *Phys. Rev.* B **53** 9579
[8] Sellier J M 2015 *J. Comput. Phys.* **297** 254
[9] Dederichs P H, Zeller R and Schroeder K 2006 Point Defects in Metals II: Dynamical Properties and Diffusion Controlled Reactions *Springer Tracts in Modern Physics* **vol 87** (Berlin: Springer)
[10] Brus L E 1984 *J. Chem. Phys.* **80** 4403
[11] Schmidt H M and Weller H 1986 *Chem. Phys. Lett.* **129** 615
[12] Pollock E L and Koch S W 1991 *J. Chem. Phys.* **94** 6766
[13] Hu Q J, Dyson N and Harlow E 1990 *EMBO J.* **9** 1147
[14] Ekimov A I 1990 *J. Lumin* **46** 83
[15] Nair G B *et al* 1987 *Appl. Environ. Microbiol.* **53** 1203
[16] Kayanuma Y 1988 *Phys. Rev.* B **38** 9797
[17] Kayanuma Y and Momiji H 1990 *Phys. Rev.* B **41** 10261
[18] Tran Thoai D B, Hu Y Z and Koch S W 1990 *Phys. Rev.* B **42** 11261
[19] Weller H 1998 *Curr. Opin. Colloid Interface Sci.* **3** 194
[20] Hu Q, Xu X, Li Z, Zhang Y, Wang J, Fu Y and Yanbin L 2014 *Biosens. Bioelectron.* **54** 64
[21] Tolbert M A, Rossi M J and Golden D M 1988 *Science* **240** 1018
[22] Tolbert M A and Middlebrook A M 1991 *J. Geophys. Res.* **96** 9371
[23] Otero M, Dittrich T, Rappich J, Heredia D A, Fungo F, Durantini E and Otero L 2015 *Electrochim. Acta* **173** 316
[24] Rorison J M 1993 *Semicond. Sci. Technol.* **8** 1470
[25] Brazhnik K, Sokolova Z, Baryshnikova M, Bilan R, Efimov A, Nabiev I and Sukhanova A 2015 *Nanomed. Nanotechnol. Biol. Med.* **11** 1065

Chapter 2

Nanoelectronics

Nanoelectronics is a branch of nanotechnology focused on the study and application of electronic components, devices, and systems at the nanometer scale. It involves manipulating materials and structures on the atomic or molecular level to create devices with enhanced performance, reduced size, and lower power consumption compared to traditional electronic devices.

Nanoelectronics aims to overcome the limitations of conventional silicon-based electronics by offering solutions for higher-density integration, faster processing speeds, and greater energy efficiency. This field holds promise for advancing various technologies, including ultra-dense memory storage, high-speed computing, flexible electronics, and advanced medical devices, driving innovation across multiple industries.

2.1 Fundamental concepts

Electronic devices that use nanotechnology (called nanoelectronics) cover a wide spectrum of materials and devices, including specific small characteristics, which are quantum mechanical properties and inter-atomic interactions needed for further analytical studies, such as advanced molecular electronics, one-dimensional nanowires and nanotubes, and hybrid semiconductors and molecular electronics. This includes 22 nm node for silicon (Si) generation of CMOS technology. Nanoelectronics are a unique technology and are different to more traditional devices.

Moore [1] noted that Si transistors are undergoing a continuous small-scale evolution, which is known as Moore's law. Since Moore made this prediction, transistor size has decreased from 10 micrometers to 28–22 nm. The goal of nanoelectronics is to continue the realization of Moore's law using new materials and methods for building nanoscale electronics. So, their volume decreases as a third power of their linear dimension, while their surface area decreases as a second power. This unavoidable and subtle principle has a big ramification. A side example for a drill's power is that it is proportional to its volume, but the drill's gear, bearing, and friction are proportional to surface area. As the length scale reduces by a factor 1000, it decreases by 1000^3 (factor of a billion), but reduces its friction by factor 1000^2 (factor of a million). So, it has 1000 times less power per friction than an original drill. If the ratio of friction-to-power is 1%, a small drill will have 10 times more power than friction—in other words, the drill is unable to do anything.

Therefore, while reduced integrated electronics are functional, the known technologies cannot be utilized for working with devices for scales where the force of friction starts to overcome the power. So, although you can watch microphotographs of etched Si gears, these devices are limited to applications such as moving shutters and mirrors [2]. In addition, at this scale the surface tension increases; thus, the tendency is to let small object stick together. This leads to an impractical 'micro factory'—if the arms and hands of robotics could be scaled down, it would be impossible for them to put down anything down after picking up it. Meanwhile, molecular development in nano-scaled muscle, flagella, and cilia fibers in aqueous environments exploit the increased forces of friction at nano- or micro-scale. Contrary to a propeller or paddle that relies on normal forces of friction (forces of friction are perpendicular to surface) to achieve pushing, cilia motion develops from laminar forces (forces of friction are parallel to surface) or exaggerated drag available at nano- and micro-dimensions. Therefore, to build a meaningful nano-machine, the relevant forces would require consideration. Consequently, we face the design and development of novel nano-machines rather than small versions of macroscopic machines. Here, nanotechnology has to be assessed for practical application.

2.1.1 Memory storage

Previously, electronic memory has been designed using large formations of transistors. Research-based electronics have presented an alternative interconnection reconfiguration between horizontal and vertical arrays to create an ultra-high-density memory. Nantero has developed a crossbar memory-based carbon nanotube (CNT) named Nano-RAM, while Hewlett–Packard has presented a memristor (specific electrical component) as a future flash memory [3].

Spintronics are included in these new devices. The resistance dependence (due to electron spin) is on an external field, which is named magnetoresistance. This effect can be amplified to what is known as giant magnetoresistance (GMR) for nanoscale objects when two ferromagnetic layers are separated by a nonmagnetic layer such as Co–Cu–Co. The GMR effect has promise for a large increase in data density storage for hard disks and may make gigabyte capacity possible. Meanwhile, tunneling magnetoresistance (TMR) such as GMR is based on electron spin tunneling within adjacent ferromagnetic layers. So, both TMR and GMR effects can be used for computers as non-volatile memory, such as magnetic random-access memory (MRAM).

2.1.2 New optoelectronics

Analog electrical devices are dramatically being replaced by optoelectronic or optical devices because of their huge capacity and bandwidth. QDs and photonic

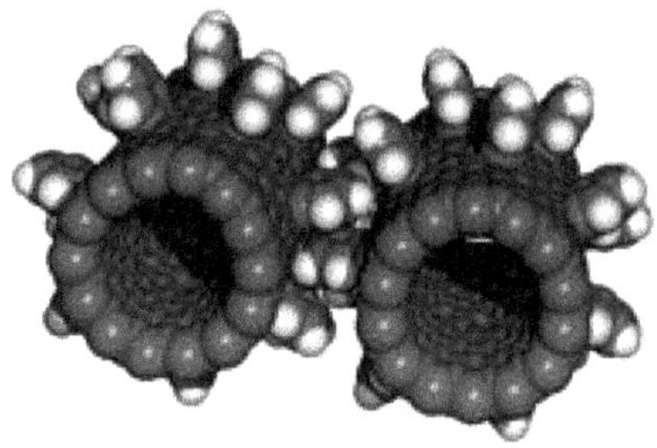

Figure 2.1. CNTs are electrically conductive and their diameters are measured in nanometers. Reprinted from [5], Copyright (2023), with permission from Elsevier.

crystals are particularly attractive and promising [4]. QDs are nano-scaled devices that are used for many things, such as laser construction. Their advantages are that they emit wavelengths depending on the QD's diameter. Laser QDs offer a higher quality and cheaper beam than a laser diode. Meanwhile, photonic crystals have periodic variations of refractive indices with respect to lattice constant using a light of half wavelength. They offer tunable bandgap for wavelength propagation that resembles semiconductors for photons or light for electrons (figure 2.1).

CNTs can be used to produce low-energy consumption devices. CNTs are conductive and because of their small size in the nanoscale it is possible to use them as highly efficient emitters for field emission displays, which relies on a cathode ray tube and very small length scale [5].

2.1.3 Production of energy

Nanostructured materials such as nanowires are an impetus for creating efficient and cheaper solar cells than traditional silicon solar cells [6]. This invention will have a great effect on the global requirements of energy, followed by producing energy devices operating in vivo, named bio-nano generators. This generator is a nano-scaled electrochemical device, such as galvanic cell or fuel cell, that draws power from blood glucose in the body, like food-based generation of energy. The enzymes that are used are able to strip glucose of its electrons, freeing them for use in electrical devices. It is known that an average body has the theoretical ability to generate 100 W of electricity (2000 food calories per day) with a bio-nano generator [7]. This is true when all of the food that is consumed is converted to electricity, the body requires some energy and this would lower the amount of power that it is possible to generate. The bio-nano generator could be embedded inside the body to generate electricity, such as sugar-fed nanorobots and pacemakers. Panasonic's Nanotechnology Research Laboratory has advanced the experimental research of bio-nano generators.

For the last five decades, research for quantum structures of low-dimensional semiconductors has been ongoing. Hence, the commercialization stage is available for optoelectronic or electronic III–V quantum well devices to achieve high-performance luminescence detectors and high-speed electronics. QDs are zero-dimensional materials, which could refer to artificial atoms. So, scientific efforts in QD research have developed applications in detectors and light-emitting devices.

QDs can also be utilized for manufacturing qubits. This means that the QD's charge can be used to define a quantum bit; '1' for one extra electron and '0' for none. Meanwhile, other researchers have different opinions on the electron's spin state (spin up and down) and consider it '1' and '0'. Current research is focusing on the QD's exciton state—when QD absorb energy of photon excitations, a self-excited electron enters a conduction band, which leaves a hole in the valence band. The hole-electron pair reduces the energy via Coulomb's attraction, which results in exciton formation. The exciton has a long lifetime, including radiation of relatively sharp lines that can easily be used for exciting and detecting via optical methods. There are optimistic opinions of QDs' potential in quantum computing. Although theoretical efforts have been made for excitement surrounding QD-based computing, no one is yet capable of testing logic operations of a quantum bit. Experimental efforts are still missing for measurement and control of QD's process.

Quantum computation covers all possible quantum phenomena in nature; therefore, it can physically realize different systems, such as electron spin, nuclear spin, superconductor, ion trap, and optical cavity [8]. Specific conditions such as quantum measurement capability, universal gates, long coherence time, initialization ability, and scalable quantum bit (qubit) system could be used for quantum computations [9]. QDs have discrete electronic states that can be regarded as localized two-level systems for constructing quantum logic gates qubits and fundamental elements. Other results of fully quantized electronic structures are semiconductor QDs to exhibit long carrier coherence time. Due to the capabilities of semiconductor manufacturing, a scalable system of quantum computing can be constructed in solid state and integrated with microelectronic technologies to satisfy the requirements of solid-state quantum gates. QDs have been researched to build quantum computers in different theoretical proposals [10] and specific experimental studies [11]. Self-assembled QDs are attractive because of the confinement properties of good homogeneity, longer coherence time, smaller size, and they are stronger carriers. To be successful for QD qubits of coherent manipulation, it is supposed to apply self-assembled QDs in quantum computers, depending on QD's fine electronic and spatial control. In the next section, semiconductor QD applications for quantum computation will be presented.

2.2 Traditional quantum dots

The history of QDs traces back to their discovery in glass crystals in 1980 by Russian physicist Alexey Ekimov [12]. Significant advances occurred in 1984 when Louis Brus established a relationship between the size and bandgap of semiconductor nanoparticles using a particle-in-a-sphere model approximation for bulk semiconductor wave functions [13]. However, it was not until nearly a decade later that QD research regained momentum following the successful preparation of colloidal CdX (X = S, Se, Te) QDs with size-tunable band-edge emissions and absorption by Murray *et al* [14]. Since then, CdX QDs have been extensively studied for their remarkable electrochemical and optical properties.

Concerns over the toxicity of cadmium ions in CdX compounds spurred efforts to enhance their biocompatibility, stability, and photoluminescence (PL) quantum yield. This led to the development of core-shell nanocrystals, where thin layers of materials with larger bandgaps encapsulate the core nanocrystals. Passivating the nanocrystals with a shell of a wide bandgap material not only improves luminescence efficiency but also prevents leaching of metal ions from the core [15]. Initially, CdSe/CdS and CdSe/ZnS were the most investigated systems [16]. Subsequently, other 'shell–core' QDs were developed, such as CdTe/CdS/ZnS 'core/shell/shell' QDs [17], CdTe/ZnS [18], CdTe/CdS [19], and CdSe/ZnSe [20]. Reiss *et al* [17] synthesized CdSe/ZnSe core/shell nanocrystals using zinc stearate as the zinc source. They synthesized CdSe core nanocrystals in a mixed TOPO/HAD solvent with a specific molar ratio, incorporating CdO complexed with dodecylphosphonic acid as the cadmium precursor. ZnO, also complexed with dodecylphosphonic acid, was injected with TOPSe into the mixture containing HAD/TOPO and CdSe core nanocrystals. After synthesis, mercaptocarboxylic acids were introduced to render them water-soluble, achieving a PL efficiency of 60%–85% in water with mercaptoundecanoic acid and in organic solvents.

CdSeTe@ZnS–SiO_2 QDs were synthesized by incorporating ZnS-like clusters into a SiO_2 shell using a microwave-assisted approach, as depicted in figure 2.2.

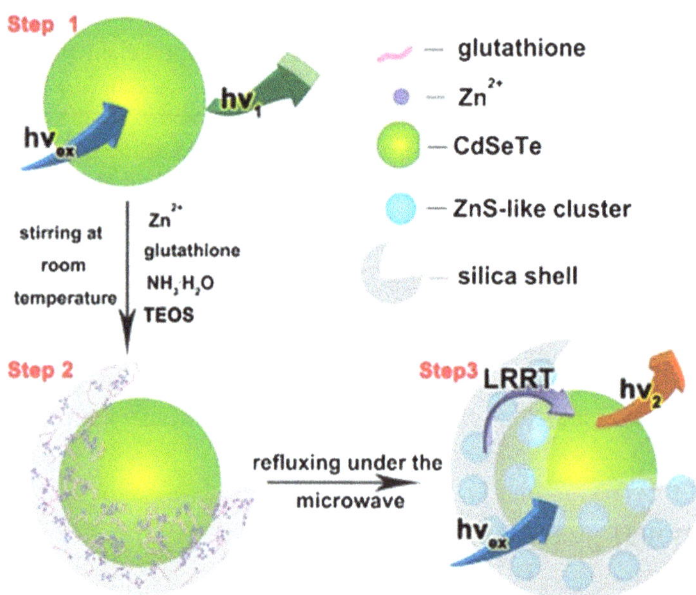

Figure 2.2. Synthetic pathway for preparation CdSeTe@ZnS–SiO_2 QDs. Reprinted from [21], Copyright (2023), with permission from Elsevier.

Green-emitting CdSeTe alloy QDs capped with mercaptopropionic acid were first synthesized and purified. Subsequently, a silica layer was coated onto the CdSeTe QDs at room temperature in the presence of Zn^{2+} and glutathione. The silica-coated QDs were then subjected to microwave irradiation, resulting in a change in fluorescence from dim green to bright orange under 365 nm excitation. The quantum yield improved from 11.9% to 56.9% using rhodamine 6 G as a standard before and after ZnS–SiO_2 coating [21].

2.3 Emerging quantum dots

Traditional QDs have predominantly been composed of cadmium, but concerns over cytotoxicity arising from leaked cadmium ions have restricted their application in cellular and in vivo studies. Consequently, there has been a growing interest in developing cadmium-free quantum dots (CFQDs) to fulfill the need for more biocompatible signal reporters in biological applications. CFQDs encompass a range of materials including silicon QDs, carbon dots, graphene QDs, Ag_2Se, Ag_2S, InP, and $CuInS_2$/ZnS. While silicon QDs have been known since as early as 1990, their excellent biocompatibility has spurred renewed interest, making them a focal point of discussion in this section. Metal nanoclusters are also emphasized for their exceptional properties and widespread use as luminescent probes in biosensing and bioimaging applications.

2.3.1 Quantum dots

Integrated circuits in QD nanophotonics are made up of nanoscale optoelectronics featuring semiconductor QDs. The invention of the transistor has revolutionized the modern electronics industry, making it an essential component in a wide range of computer and communication system applications [22].

Since the early-1980s, QDs have gained increasing attention and visibility. With advances in modern materials science techniques, the focus has shifted from modeling their behavior to implementing and leveraging their unique properties for various applications. Initially conceived as the next logical step in lasing devices after quantum well and quantum film structures, QDs progressed from one- to two- and then to three-dimensional confinement of electron and hole wave functions. This progression narrows the gain spectrum, increases maximum gain, and reduces the threshold current required to initiate laser operation. Additionally, the quantization of electron and hole energy states, along with material and size-specific absorption characteristics, enables high extinction ratios of QD optical labels in biomedical research. These properties also form the basis for various photonic devices, including transistors, switches, and lasers.

Problems

1. What is nanoelectronics? And, how is it different from traditional electronics?
2. How do nanoelectronics devices such as nano-transistors and nanowires function? And, what are their potential applications?

3. What are some of the key challenges involved in manufacturing nanoelectronics devices? And, how are researchers working to overcome these challenges?
4. How do quantum effects come into play in nanoelectronics? And, how do they affect the behavior of devices at the nanoscale?
5. How are nanoelectronics being used in emerging fields such as quantum computing, neuromorphic computing, and energy harvesting?

References

[1] Melosh N, Boukai A, Diana F, Gerardot B, Badolato A, Petroff P and Heath J R 2003 *Science* **300** 112

[2] Kittel C 1979 *Introduction to Solid State Physics* 5th edn (New York: Wiley)

[3] Goicoechea J, Zamarreñoa C R, Matiasa I R and Arregui F J 2007 *Sens. Actuators* B **126** 41

[4] Waldner J-B 2007 *Nanocomputers and Swarm Intelligence* (London: Wiley)

[5] Cao M, Wang L, Zhang Q, Zhang H, Zhong S and Chen J 2023 *Mater. Today Commun.* **36** 106677

[6] Tian B, Zheng X, Kempa T J, Fang Y P, Nanfang Y, Guihua H J and Lieber C M 2007 *Nature* **449** 885

[7] Jiang X, Huang W and Zhang S 2013 *Nano Energy* **2** 1079

[8] Nielsen M A and Chuang I L 2000 *Quantum Computation and Quantum Information* (Cambridge: Cambridge University Press)

[9] DiVincenzo D P 2002 *Semiconductor Spintronics and Quantum Computation* ed D D Awschalom, D Loss and N Samarth (Berlin: Springer)

[10] Lovett B W, Reina J H, Nazir A and Briggs G A D 2003 *Phys. Rev.* B **68** 205319

[11] Li X, Wu Y, Steel D, Gammon D, Stievater T H, Katzer D S, Park D, Piermarocchi C and Sham L J 2003 *Science* **301** 809

[12] Ekimov A I and Onushchenko A A 1981 *JETP Lett.* **34** 345

[13] Brus L 1986 *J. Phys. Chem.* **90** 2555

[14] Murray C B, Norris D J and Bawendi M G 1993 *J. Am. Chem. Soc.* **115** 8706

[15] Dabbousi B O, Rodriguez-Viejo J, Mikulec F V, Heine J R, Mattoussi. H, Ober R, Jensen K F and Bawendi M G 1997 *J. Phys. Chem.* B **101** 9463

[16] Peng X, Schlamp M C, Kadavanich A V and Alivisatos A P 1997 *J. Am. Chem. Soc.* **119** 7019

[17] Reiss P, Bleuse J and Pron A 2002 *Nano Lett.* **2** 781

[18] He Y, Lu H T and Sai M 2006 *J. Phys. Chem.* B **110** 13370

[19] Zhao D, He Z, Chan W H and Choi M M F 2009 *J. Phys. Chem.* C **113** 1293

[20] He Y, Lu H T, Sai L M, Su Y Y, Hu M, Fan C H, Huang W and Wang L H 2008 *Adv. Mater.* **20** 3416

[21] Sapna K, Manzur Ali P P and Mohamed Hatha A A 2023 Nanomaterials in bioimaging and cell labeling ed S Thomas, N Kalarikkal and A R Abraham *Applications of Multifunctional Nanomaterials* (Amsterdam: Elsevier) p 499

[22] Zhu J-J, Li J-J, Huang H-P and Cheng F-F 2013 Quantum Dots for DNA Biosensing *Springer Briefs in Molecular Science* (Berlin: Springer) doi: 10.1007/978-3-642-44910-9

IOP Publishing

Quantum Dots
Synthesis, characterization, and optical investigations
Yarub Al-Douri

Chapter 3

Synthesis of quantum dots

The synthesis of quantum dots (QDs) involves creating nanoscale semiconductor particles that exhibit unique optical and electronic properties due to quantum confinement effects. QDs typically range from 2 to 10 nm in diameter and have size-tunable properties, making them useful in a wide range of applications, such as bioimaging, photovoltaics, light-emitting devices, and quantum computing.

Various methods exist for synthesizing QDs, including colloidal synthesis, epitaxial growth, and template-assisted methods. Colloidal synthesis is the most common approach, where QDs are formed in a solution containing precursors, surfactants, and solvents. This method allows precise control over the size, shape, and surface chemistry of the QDs by adjusting reaction parameters such as temperature, time, and precursor concentrations.

3.1 Characteristics of quantum dots

Nanocrystals (NCs), including semiconductor particles and monodisperse crystalline metals, come in various forms, such as QDs with adjustable diameters at the nanoscale. QDs are particularly interesting for devices due to their unique photon and electron behaviors, which differ significantly from those in microscale (bulk) materials. When QDs are deposited in a solid, ordered manner, their collective properties become distinctive. The performance of individual QDs can be tuned by varying their shape and size, affecting hole–electron recombination and generation dynamics, energy transfer, and tunneling behavior. In bulk materials, population inversion can be achieved with sufficient energy and excitation rates, resulting in an emission coefficient greater than the absorption coefficient, leading to net optical gain. Additionally, the three-dimensional confinement of hole–electron pairs (exciton wavefunctions) in QDs results in discretized energy levels and delta-function-like density of states (DoS). Under excitation, QDs exhibit sharper emission compared to their bulk counterparts, resulting in sharp optical gain, high quantum efficiency, low loss, and high gain (figure 3.1) [1]. The optical response of QDs can be tuned by

doi:10.1088/978-0-7503-5704-3ch3

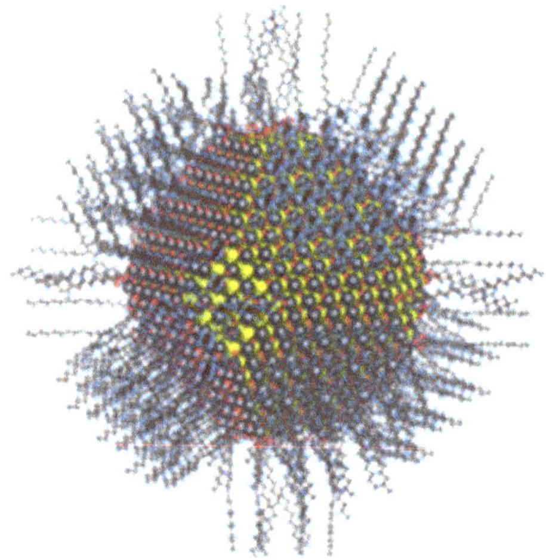

Figure 3.1. Colloidal lead sulfide (selenide) nanoparticle with complete passivation by hydroxyl (size ~5 nm), oleyl, and oleic acid. Reprinted from [5], Copyright (2023), with permission from Elsevier.

their composition and size, making them ideal for optoelectronics due to these unique properties.

QD photonics utilize self-organization at the strained interface of epitaxially grown II–VI or III–V semiconductors to create pyramidal QD nanostructures due to lattice mismatch. MOCVD is employed for fabrication. The Stranski–Krastanov process has been used to manufacture the first QD laser, demonstrating improved gain characteristics and reduced laser current thresholds compared to quantum film and quantum well structures. Despite the high-quality optoelectronics achieved through the self-organization method, it is not compatible with silicon CMOS fabrication. Solution-based processes and chemical synthesis are alternative methods for QD fabrication, resulting in colloidal QDs, which are widely used as fluorescent tags in biomedicine. These colloidal QDs can be deposited on various substrates, including silicon, using different chemical methods, leading to applications such as solar cells, organic LEDs, sensors, and whispering gallery lasers. For individual colloidal nanoparticles fabricated in solution suspension, semiconductor materials from groups III–V and II–VI are used. The process involves controlled reagent injection into a solvent for compound nucleation and growth. Research has also been conducted on the preparation of group-IV QDs.

For CdSe NCs, organophosphine selenide and dimethylcadmium combine to create the core QD structure, while a shell layer, such as ZnS, may be added by solution nucleation to prevent surface oxidation and improve fluorescence. The core diameter and shell thickness determine the QDs' energy levels, affecting the absorption spectrum and corresponding photon emission wavelength during electron-hole recombination and excitation [2]. The shell layer allows for further chemical surface treatment to enable reactions with surrounding mediators. Designing a shell with a larger bandgap provides better confinement of electron-hole pairs, and the surface can be treated with stabilizers conjugated with molecules of specific binding activity depending on the application. Colloidal QDs offer high integration flexibility and ease of fabrication. The next section will focus on photonics deposition and fabrication using colloidal QDs.

3.1.1 Synthesis of colloidal nanocrystals

To synthesize colloidal NCs, chemical processes are used to dissolve precursor compounds in specific solutions. These precursors are utilized to synthesize solvents, organic surfactants [3], and colloidal QDs [4]. The solution is heated to increase the temperature, causing the precursor to decompose and form monomers that nucleate to generate NCs. Temperature is a critical parameter in determining the optimal conditions for NC growth. It must be high enough to allow for the annealing and rearrangement of atoms during synthesis, but low enough to promote controlled growth of the NCs. Monomer concentration is another crucial parameter in NC growth. There are two different growth behaviors: 'focusing' and 'defocusing'. At high monomer concentrations, the critical size, where NCs neither grow nor shrink, is small, resulting in all particles growing. In this focusing behavior, small particles grow faster than larger ones, leading to monodisperse particles. When the monomer concentration is maintained so that the average NC size is larger than the critical size, the size distribution focuses optimally. As the monomer concentration decreases, the critical size becomes larger than the average size, causing the size distribution to defocus.

Various methods of colloidal synthesis are available for different semiconductors. QDs can be synthesized from binary compounds such as indium phosphide, indium arsenide, cadmium sulfide, cadmium selenide, lead selenide, and lead sulfide. They can also be made from ternary alloys such as cadmium selenide sulfide. QDs typically contain about 100–100 000 atoms within a specific volume, with diameters ranging from 2 to 10 nm (corresponding to 10–50 atoms). To put it into perspective, approximately 3 million QDs with a diameter of 10 nm would fit end to end across the width of a human thumb.

Large batches of QDs may be prepared using colloidal synthesis. Because of convenience and scalability of benchtop conditions, the methods of synthesized colloidal are promising for commercialization.

3.1.2 Fabrication of quantum dots

The fabrication of QD solutions involves methods for achieving colloids in controlled behavior [3]. First, the solvent contain reacting chemicals and stabilizing is poured into specific vessel and is then heated to a specific temperature for forming NCs. The solution is supersaturated by rapid injection of metal-organic materials, and then breaks above threshold of nucleation. Thereafter, small particles are formed and nucleate to reduce the energy of the system. Furthermore, if the injection rate is below the reagent's rates of consumption, it is expected that new nuclei will be created and existing NCs will seem to be larger with metal-organic precursors.

For a specific compound, a long chain of alkane attached to phosphorous oxide or phosphorous is used to provide active ingredients in a mixture of solvents.

To fabricate II–VI NCs, the precursors include metal alkyls that consist of bistrimethylsilyl chalcogenides incorporate group-VI, organophosphine chalcogenides, and group-II metals. For CdSe QDs, organophosphine selenide or dimethylcadmium are utilized.

To ensure monodisperse NCs, the distribution of size will give a homogeneity where initial nucleation produces particles, in which the eventual dimension eclipses an original volume. Furthermore, the Ostwald ripening is a secondary phase of growth that dissolves smaller NCs with high energy of surface for deposition on a larger one. In general, the process or procedure is timed or calibrated to synthesize particles of a required size. In addition, the NCs and reactants are omitted from the solution.

To create shell/core QDs, the material has compatible conditions of deposition and energy of surface energies at the interface. Another condition is that the threshold of nucleation for shell must be less than the core, and the interdiffusion between shell and core elements should not occur. The bilayer QD production method appends to core procedure of synthesis by choosing the required size NCs for adding them to a new reaction solution. For optimum temperature, the shell precursor and material are added at a specific rate to permit a gradual deposition onto the original particle while avoiding secondary nucleation [7]. For the final capping layer, the exchange process is utilized via stabilized molecules attached to NCs that are interchanged within competition with more other molecules, ions, and molecule groups. Moreover, clumping of the particle is avoided when the strength of repulsion for the outermost layer is larger than the van der Waals attractive force.

Generally, to produce core/shell and core structures, the size should have a 5–10% standard deviation of the mean. More separation and filtration steps may be achieved to collect and define NCs of best uniformity. Some standard methods of structural analysis and characterization of NCs include transmission electron microscopy and both wide- and small-x-ray scattering, while the chemical constitutions could be analyzed via photoelectron spectroscopy, x-ray fluorescence, and nuclear magnetic resonance imaging.

Since its cost-effective and nontoxic, the Si QD is an active topic and efforts are invested to develop its technologies for future commercialization. The synthesis development of colloidal Si QD has delayed beyond II–VI QDs researches; so, the process of Si QD including attractive characteristics has been investigated. Hydrogen-terminated colloidal Si QD can be prepared utilizing the low-temperature solution process [7].

Si QDs can be produced utilizing the electrochemical etching technique [8]. The process resumes with doping p-type boron with Si wafer of orientation (100) and 5–20 Ω-cm resistivity in a mixture of polyoxometalates and hydrogen peroxide (H_2O_2) catalysts, and methanol and hydrofluidic acid. At several hours of etching time and high current density, $>10\ \mathrm{mA\ cm^{-2}}$, micropores of diameter, 2 nm arise as typical structures, while nano pores are formed at short etching time and low current density, $<10\ \mathrm{mA\ cm^{-2}}$. The ultrasonication technique is utilized to disintegrate the etched wafer into colloidal Si QDs. But before that the reaction of hydrosilylation with chloroplatinic acid as catalyst results in octane-modified Si surface to prevent aggregation of the QDs.

3.1.3 Quantum confinement effect

For NCs, the excitons are squeezed when the radius is smaller than their exciton Bohr radius, known as quantum confinement. The levels of energy can be modeled utilizing the box model of the particle, where different energy states depend on box length. The QDs are in a 'weak confinement regime' if their radii are on the exciton Bohr radius. Meanwhile, QDs are in a 'strong confinement regime' if their radii are smaller than exciton Bohr radius. If the QDs' size is less than 10 nm, the quantum confinement effect dominates, and optical and electronic properties are largely tunable.

The splitting of energy levels for small QDs is due to quantum confinement effect. The horizontal axis is the radius or size of QDs and a_b* is the exciton Bohr radius. Fluorescence happens if excited electrons relax to ground state and combine with holes. Simply, the emitted photon energy can be consumed as a summation of energy bandgaps between occupied and unoccupied levels, bound energy of exciton (hole–electron pair), and energies of confinement for holes and excited electrons.

Since to the electrons are limited in movement along specific directions, the energy of electron in the mentioned direction could be quantized, which is attributed to its direction, and bound electrons could form standing waves. Based on the binding dimension number, materials can be classified into zero-dimension (bulk), one-dimension (quantum well), two-dimension (quantum wire) and three-dimension (QD). The quantization of energy has a great impact in terms of electronic density DoS, defined as:

$$\mathrm{DoS} = \frac{dN}{dE} = \frac{dN}{dk}\frac{dk}{dE} \tag{3.1}$$

The quantum confinement is a reason for discrete states, which can be obtained with the energy-level solution of Schrodinger's equation:

$$-\frac{\hbar^2}{2m}\nabla^2 \psi + V(r)\psi = E\psi \tag{3.2}$$

For the case of one-dimension infinite square potential well, its solution is

$$\psi(\chi) \approx \sin\left(\frac{n\pi\chi}{L}\right) \tag{3.3}$$

where integer refers to n. Figure 3.2 illustrates ground-state wave function. If the restriction is applied in one direction like x-direction, the other two directions (y, z) of energy are continuous. The total energy is (figure 3.3)

$$\frac{n^2h^2}{8mL^2} + \frac{p_y^2}{2m} + \frac{p_z^2}{2m} \tag{3.4}$$

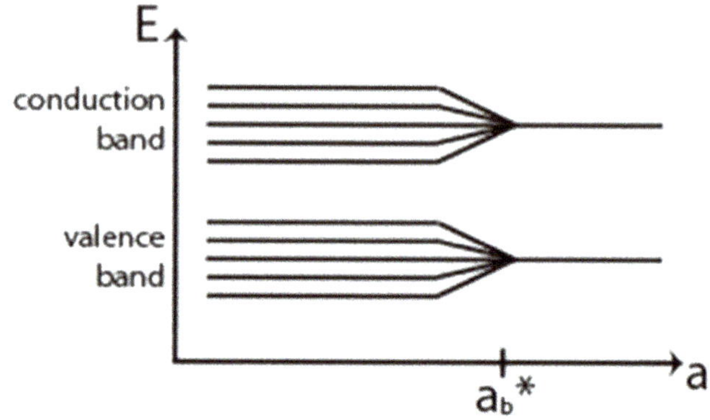

Figure 3.2. Splitting of energy levels for small QDs due to the quantum confinement effect.

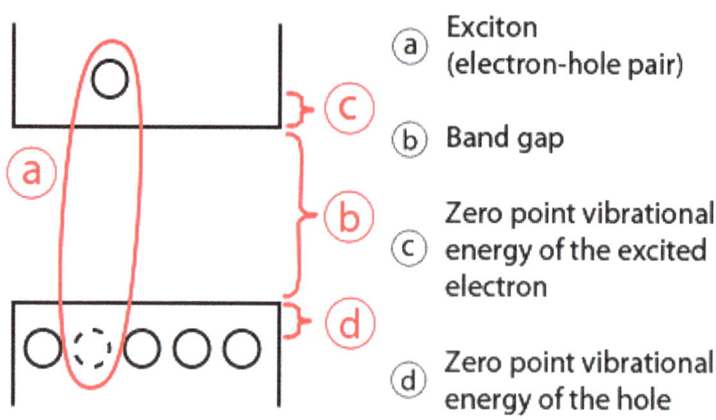

Figure 3.3. Simplified representation showing the excited electron and the hole in an exciton entity and the corresponding energy levels.

Meanwhile, for 3D infinitely deep square potential well (quantum box),

$$\psi(x, y, z) \approx \sin\left(\frac{n\pi x}{L_x}\right)\sin\left(\frac{n\pi y}{L_y}\right)\sin\left(\frac{n\pi z}{L_z}\right) \tag{3.5}$$

Here integers refer to n, m, and q. The energy level is;

$$E_{n,\,m,\,q} = \frac{n^2h^2}{8mL_x^2} + \frac{n^2h^2}{8mL_y^2} + \frac{n^2h^2}{8mL_z^2} \tag{3.6}$$

Rather than an infinite box depth, the actual barrier of potential can be restricted, spherical or present harmonic potential limitations of an oscillator, which is considered as one electron in this case. Meanwhile, for an actual situation, it is required to accord with multiple-particle and hole–electron pairs. The particle mass requires consideration at the particles' boundaries, along with a mismatch of potential. The QDs' size is approximately 10–100 nm, which is equivalent to the de Broglie wavelength. Hence, the hole or electron would be subjected to three-dimensional quantum confinement effect to have quantized levels of energy forming a zero-dimensional electronic system. The QDs are referred to as artificial atoms because QDs have electronic configuration, which is very similar to atoms, and so the QDs levels of energy are presented with symbols: *s*, *p*, and *d* refer to ground state of QDs and excited state level. For InAs QDs, their geometric shape is like a flat convex lens; so, the energy of potential in direction *r* can be approximated with a two-dimensional paraboloid. This tells us that the ground and excited states have specific numbers of degeneracies, 2, 4, 6, etc (including spin degeneracy). Meanwhile, QDs are filled with two electrons, energy level *p* is filled by four electrons, and so on. QDs do not have the same electronic configurations as atoms [5, 6].

For the orbits *s*, *p*, and *d*, the atomic levels of energy can be filled with 2, 6, and 10 electrons. The difference is that QDs have a different form of potential energy to atoms. The atomic potential is formed in Coulomb interaction, with three-dimensional symmetry. Anyway, the QDs' potential energy form is related to geometric shape. Since their shape is like a convex lens, their height is smaller than their diameter. So, the QDs electrons have two-dimensional symmetry (plane) with energy-level degeneracy form, unlike atoms.

3.1.4 Luminescence and excitons

Excitons are formed when holes and electrons come together under the influence of Coulomb forces. In semiconductors, holes and electrons are attracted to each other due to their opposite charges, with holes having an effective mass larger than that of electrons, resembling a hydrogen-like atomic system. The binding energy of excitons is defined by Bohr theory,

$$E_n = -\frac{e^2}{2\varepsilon a_0 n^2};\quad a_0 = \frac{\varepsilon h^2}{4\pi^2 \mu e^2}\quad \text{here } \mu \text{ is reduced mass.} \tag{3.7}$$

In QDs, excitons arise and are subjected to specific size constraints imposed by the QDs. Excitons exhibit discrete energy levels, resulting in absorption spectra characterized by peaks similar to a delta function δ in an exciton absorption spectrum [7].

3.1.5 Exciton's energy band structure

The semiconductor band structure consists of filled valence and unfilled conduction bands at zero temperature, separated by an energy bandgap. In cubic lattice symmetry, the top of the valence band corresponds to a state with angular momentum ($l = 1$).

Heavy and light holes are categorized based on their effective masses in the vertical direction. Due to the constraints imposed by the vertical dimension, QDs typically have a greater lateral width than vertical height, emphasizing strong vertical confinement. Light holes are lower in energy due to their smaller effective mass. Therefore, near the energy bandgap, light absorption primarily arises from contributions of heavy holes [8].

When electrons are excited from the valence to the conduction band, they leave behind positively charged holes. These electrons and holes combine to form excitons due to Coulomb attraction, releasing a binding energy of 6 meV upon formation. In two-dimensional quantum wells, the confinement effect increases the exciton binding energy by 15 meV, which is attributed to the reduced distance between positive and negative charges, thereby strengthening the Coulomb force. In the case of QDs, the exciton binding energy is further increased to 20 meV.

The energy of the exciton band equals the energy bandgap (E_g), adjusted for the binding energies of the electron E_e and hole E_h, with the exciton binding energy E_b subtracted:

$$E_{\text{ex}} = E_g + E_e + E_h - E_b \tag{3.8}$$

When an exciton recombines, a photon is emitted with energy equal to the difference between the conduction and valence bands. Excitons can have long lifetimes, up to nanoseconds (10^{29} s), making them observable with experimental techniques that enable the study of their quantum evolution. As the intensity of incident light increases to generate hole–electron pairs, multi-exciton states become necessary.

3.1.6 Synthesized semiconductor quantum dots

QDs exhibit heterogeneous nanostructures, typically coated with a material of lower energy bandgap compared to their cores. The methods used for synthesizing QDs are categorized as follows [9]:

1. Chemical colloidal method: Also known as the sol–gel method, this technique is straightforward and suitable for producing multilayer QDs. It is favored for its simplicity and potential for mass production.
2. Self-assembly method: Utilizing processes such as chemical vapor deposition (CVD) or MBE, this method leverages lattice mismatch principles to induce the self-polymerization of QDs on specific substrates. It is effective for achieving mass production and uniform arrangement of QDs.
3. Lithography and etching: Involves direct application of techniques such as electron beam lithography on substrates, which are then etched to produce patterned QDs. This method is time-consuming and not suitable for mass production due to its limitations.
4. Split-gate approach: In this method, a voltage is applied to create two-dimensional confinement within a quantum well's plane. The gate allows for control over the size and shape of QDs, making it more suitable for research purposes than for mass production applications.

These synthesis methods enable the fabrication of QDs with tailored properties and structures, catering to diverse applications in fields such as optoelectronics, photonics, and biomedical research.

The synthesis of QDs is based on the growth processes similar to thin films. According to the literature, there are three traditional models describing thin film growth:

1. Growth of layer (Frank–van der Merwe (F–M)) mode:
   The F–M model begins with a single layer deposition followed by subsequent layer-by-layer growth. The crystalline orientation of the thin film is influenced by the crystalline directions of the initial layer. In heteroepitaxy, where there may be lattice mismatch between the substrate and the growing film, and thin film stress can result.
2. Growth of island (Volmer–Weber (V–W)) mode:
   In the V–W mode, atoms are deposited onto the substrate surface, forming small islands that gradually increase in size or decompose into separate atoms. Before the formation of thin films, there is often a rearrangement over a wide range, leading to coarsening. The growth direction of the crystalline structure cannot be precisely determined during this process.
3. Mixed growth (Stranski–Krastanov (S–K)) mode:
   The S–K model combines aspects of both layer and island growth. Initially, a layer begins to grow, and then islands form either on top of a single layer or on several grown layers. This mixed mode of growth results in structures where islands coexist with continuous layers.

These growth models are crucial for understanding and controlling the formation of QDs, particularly in terms of their size, shape, and crystalline properties, which are essential for tailoring QDs for various applications in electronics, photonics, and biological imaging.

The synthesis of QDs with exceptional optical properties using self-organization techniques is a significant focus in QD synthesis efforts, aligning well with traditional technologies. Self-organization in QD growth can be achieved through methods such as MOCVD or MBE. These techniques enable the formation of three-dimensional islands (QDs) between materials with different lattice structures, even when there is a mismatch between the growth patterns of these materials. This approach is crucial for tailoring QDs with specific sizes, shapes, and optical characteristics suitable for diverse technological applications.

An illustrative example of indium arsenide QD growth exemplifies the synthesis technology. Utilizing MBE and the Stranski–Krastanov (S–K) growth mode, the process begins with a substrate of *n*-doped GaAs. Plasma hydrogen-ion treatment is applied at 610 °C to remove surface oxides, followed by the growth of a buffer layer several hundred nanometers thick. The temperature is then adjusted to 500 °C (range: 450 °C–550 °C) for epitaxial growth of indium arsenide (InAs), with the layer thickness increasing with generation time and rate.

Self-assembled quantum dots (SAQDs) of InAs begin to form once the thickness exceeds 1.8 monolayers (ML), marking a threshold thickness. The quality of SAQDs in terms of morphology, uniformity, density, and size is intricately linked to parameter settings, growth conditions, and methods employed. The high-quality growth of SAQDs represents a cutting-edge technology.

Encapsulation of these QDs with a GaAs layer facilitates their use in devices such as the active layer in laser diodes. To confine carriers effectively for optical purposes, resonant cavities are created by growing barrier layers on both sides of the active layer, using different *n* (or *p*) concentrations of AsAlGa alloy. The outermost layers, *p*1 (or *n*1), highly doped, serve for forming ohmic contacts with electrodes.

Problems

1. What are QDs? And, how are they synthesized?
2. What are the most commonly used methods for synthesizing QDs? And, what are their advantages and disadvantages?
3. How can the size and shape of QDs be controlled during synthesis? And, why is this important for their properties and applications?
4. What are some of the challenges involved in synthesizing high-quality QDs? And, how are researchers working to overcome these challenges?
5. What are some potential applications of QDs? And, how does their synthesis impact these applications?

References

[1] Xu R, Jiang Y, Xia L, Zhang T, Xu L, Zhang S, Liu D and Song H 2015 *Biosens. Bioelectron.* **74** 411

[2] Asada M, Miyamoto Y and Suematsu Y 1986 *IEEE J. Quantum Electron.* **22** 1915

[3] Bruchez M, Moronne M, Gin P, Weiss S and Alivisatos A P 1998 *Science* **281** 2013

[4] Murray C B, Kagan C R and Bawendi M G 2000 *Annu. Rev. Mater. Sci.* **30** 545

[5] Sapna K, Manzur Ali P P and Mohamed Hatha A A 2023 Nanomaterials in bioimaging and cell labeling ed S Thomas, N Kalarikkal and A R Abraham *Applications of Multifunctional Nanomaterials* (Amsterdam: Elsevier) p 499

[6] Shi D, Guo Z and Bedford N 2015 *Nanomaterials and Devices* (Amsterdam: Elsevier)

[7] Neiner D, Chiu H W and Kauzlarich S M 2006 *J. Am. Chem. Soc.* **128** 11016

[8] Tu C-C, Tang L, Huang J, Voutsas A and Lin L Y 2010 *Opt. Express* **18** 21622

[9] Zhan G Z 2003 *Physics* **25** 1

Chapter 4

Graphene and quantum dots

Graphene and quantum dots (QDs) represent two frontier materials in nanotechnology, each with distinct and remarkable properties. Graphene has a single layer of carbon atoms arranged in a two-dimensional honeycomb lattice exhibiting extraordinary properties. Its unique electronic properties, such as high carrier mobility and tunable bandgap, make it a promising material for applications in transistors, sensors, energy storage devices, and flexible electronics.

QDs, on the other hand, are nanoscale semiconductor particles that exhibit size-dependent optical and electronic properties due to quantum confinement effects. These properties include discrete energy levels and tunable emission spectra, which make QDs highly suitable for applications in bioimaging, light-emitting diodes (LEDs), solar cells, and quantum computing.

QDs exhibit size-dependent optical and electronic properties due to quantum confinement effect. These distinguished properties include discrete energy levels and tunable emission spectra, which make QDs highly suitable and appropriate for bioimaging, light-emitting diodes (LEDs), solar cells and quantum computing applications.

4.1 Nanostructured graphene and quantum dots

This section will describe the synthesis methods, followed by experiments for nanostructured graphene and QDs, with a focus on size and edge quantization effect. Major interests in nanostructured graphene are related to electronic applications, such as photodetectors, transparent electrodes, and transistors [1]. To switch transistors on and off, the energy bandgap is necessary to regulate current. Graphene is semiconducting material with almost zero energy bandgap and lowest conductivity at point Dirac, where current could not switch off.

In addition, as a result of the Klein paradox, it is not easy to seize electrons via an electro-static gate. To reduce graphene's lateral size, it is easy to solve zero energy bandgap problem. So, for the size quantization, the energy bandgap opens.

doi:10.1088/978-0-7503-5704-3ch4

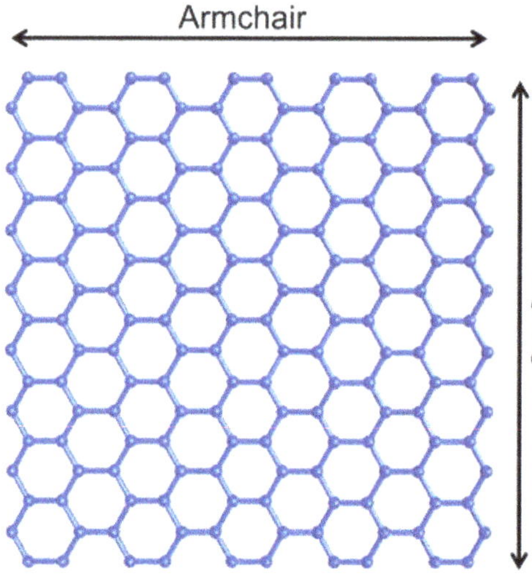

Figure 4.1. Two possible edges termination of graphene quantum dots. Reprinted from [2], Copyright (2023), with permission from Elsevier.

Meanwhile, the finite-size of semi-metallic graphene makes it a semiconducting material. For nanostructured graphene, graphene QDs (islands) and graphene ribbons (strips) are interesting. Cutting nanostructured graphene out of graphene leads to two different edges: zigzag and armchair, as shown in figure 4.1. Nanostructured graphene may be characterized by whether the symmetry of the sublattice is conserved or not. Therefore, both types of edges, i.e., absence and presence of symmetry of sublattice, have an important role in defining the electronic properties of nanostructured graphene.

4.1.1 Methods of fabrication

Graphene can be modelled into ribbons (graphene nanoribbons; GNRs) of different widths using etching mask and electron-beam lithography, as suggested by [2]. Mechanical exfoliation is employed to obtain high-quality nanostructured graphene. So, graphene is deposited onto substrate of *p*-doped Si covered by a layer of SiO_2. The nanostructured graphene strips are covered by a protective etch mask made of molecules of cubical-shaped that have one Si atom at each corner and are linked with corners by atoms of oxygen and hydrogen to form silsesquioxane (HSQ). The unprotected nanostructured graphene is etched by a plasma of oxygen. Using this specific technique, Han *et al* [2] were able to do measurements of transport on 20–500 nm width and $\sim$1 μm length samples. They noted that the transport properties depend on both trapped charges and boundary scattering of substrate.

Jia *et al* [3] proposed a different method for creating ribbons. They utilized electron-beam irradiation and Joule heating [4]. Where the samples are exposed to irradiation of electrons for 20 min and heated by directional high-electrical current. During the heating process, the carbon atoms on the sharp edges are evaporated and GNRs of smooth edges are created. Meanwhile, Li *et al* [5] have chemically built nanostructured graphene nanoribbons with clarified edges. The ribbons' width are ~10–50 nm and their length is ~1 μm. They reported nanostructured graphene of irregular shapes and observed the ribbons of 120° zigzag and kink edges. Meanwhile, previous work is studied with the thinnest ribbon, ~10 nm width. Cai *et al* [6] have suggested a technique for creating a ribbon of width <~1 nm. They resumed of colligated monomers that define the ribbon's width. These monomers are deposited onto net substrate surfaces by sublimation of a sixfold evaporator. They utilized two steps of annealing processes of different temperatures for chevron-type ribbons. Different chemical approaches for creating nanostructured graphene of different shapes are suggested [7]. The many shapes imply different chiralities of graphene nanoribbon. The chirality is conned to an angle where ribbon is cut. The GNRs have many widths and chiralities that are chemically prepared by unzipped CNTs [8]. The presence of a one-dimensional GNR edge is affirmed via STM. The comparison of theoretical prediction and experimental results are based on density functional theory (DFT) methods and the Hubbard model to provide evidence for spin-polarized edge state formation [9]. It is seen that magnetic and electronic properties can be tuned by changing the width and edge chirality [10]. Partial unzipped CNTs have been investigated [10]. Topology defects like the interface between two nanostructured graphenes. The emergence of spatial localized interface state is predicted [11] and its rules for edge states existence are elaborated [12].

Nanostructured graphene nanoribbons have two-dimensions limited to one-dimension, while QDs are two-dimensional systems restricted to two directions. Chemistry supplies a normal way to develop nanostructured graphene QDs that have hundreds of atoms. Müllen *et al* [13] have utilized the bottom-up approach of molecular nanostructured graphene to unconventional carbon and a synthetic way to process and tailor nanostructured graphene on a metal surface. Meanwhile, Li *et al* [14] have promoted a chemical way to develop colloidal nanostructured graphene QDs including up to 200 atoms of carbon, as illustrated in figure 4.2(a). Ajayan *et al* [15] used oxidation cutting and graphitic fibers to manufacture nanostructured graphene QDs of different shapes, as illustrated in figure 4.2(b). Berry *et al* [16] have improved nano-anatomy (nanotomy)-based dispersible and transferable nanostructured graphene production of controlled size and shape. Such methods are required if nanostructured graphene QDs are utilized for wide energy applications, as reviewed by Zhang *et al* [17].

For optical electronic applications, QDs of different sizes exceeding those prepared via bottom-up approach and full control over edge and shape are required. So, the top-down technique is included using atomic force microscopy (AFM). Bunch *et al* [18] have attempted to top-down fabricate nanostructured graphene QDs. They researched graphite QDs of thickness of a few tens of nanometers and lateral dimension ~1 μm. They deposited onto Si wafer, 200 nm, grown oxide thermally and connected to metallic electrode. The measurements of transport properties have shown Coulomb blockade phenomenon. To analyze Coulomb oscillation period in voltage gate, they have demonstrated QDs area extended into

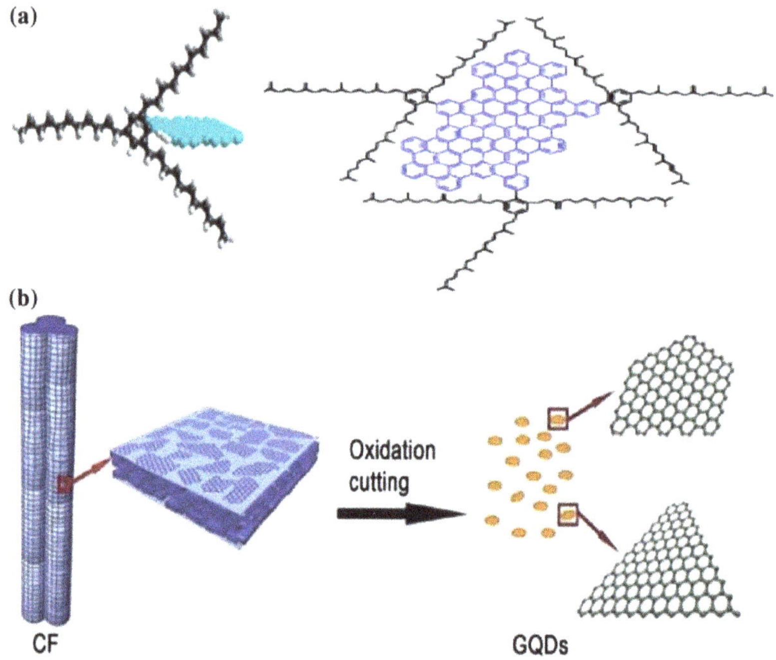

Figure 4.2. (a) Colloidal nanostructured graphene QDs. (b) QDs of graphitic fibers by oxidation cutting. Reprinted with permission from. Reprinted from [2], Copyright (2023), with permission from Elsevier.

a piece of graphite under electrodes. Nanostructured graphene QDs are experimentally manufactured from graphene sheets. Ponomarenko *et al* [19] have made different nanostructures with plasma of oxygen etching and protecting mask utilizing high-resolution electron-beam lithography. Their technique has allowed us to create QDs of 10 nm radius without a well-defined shape. On the other hand, Liu *et al* [20] have researched tunable nanostructured graphene QDs manufactured with reactive ion etching modelled graphene, as illustrated in figure 4.3(a). Pereira *et al* [21] have predicted bilayer graphene, where the inhomogeneous gate applications on top of bilayer graphene allow the charge carriers to be confined, as shown in figure 4.3(b).

The alternative fabrication method is to create nanostructured graphene via cutting graphene into specific shapes. It is seen that few-layer [22] and single-layer [23] nanostructured graphene can be cut with metallic particles. This process is based on anisotropic etching by thermally activated nickel particles. The cuts are directed along proper orientations of crystallography, with width determined by the diameter of the metal particles. Using this technique, the authors are able to make nanostructured graphene, equilateral triangles, and ribbons. Another technique to fabricate nanostructured graphene is to use direct growth on metallic surfaces, and is characterized by AFM [24]. Another example of triangular graphene QDs grown on an Ni surface are illustrated in figure 4.4(a) [25], nanostructured graphene

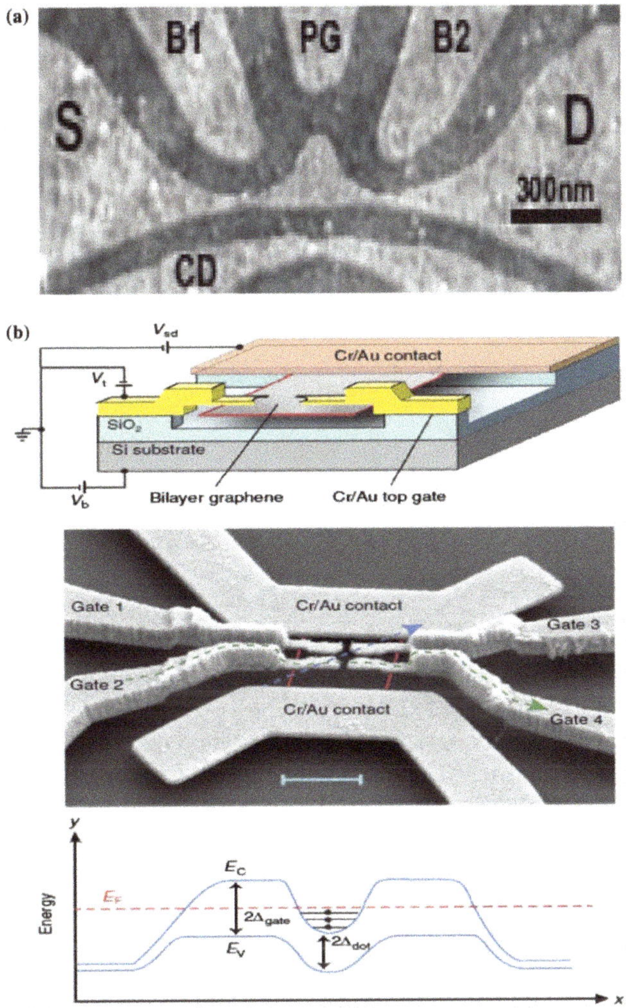

Figure 4.3. SEM picture of (a) QD etched out of graphene, and (b) QD defined by gates in bilayer graphene. Reprinted from [2], Copyright (2023), with permission from Elsevier.

QDs on an Ir surface are shown in figure 4.4(b) [26], and graphene QDs on a Cu surface are shown in figure 4.4(c) [27].

4.2 The edge's role

Figure 4.1 illustrates a honeycomb lattice that has two distinct edges: zigzag and armchair. They are experimentally noticed close to the single-step edge on exfoliated graphite surface via spectroscopy characterization, scanning tunneling microscopy (STM) [28], and Raman spectroscopy analysis [29]. The armchair and zigzag edges are investigated by various activation energies. The estimated calculations of activation energies for molecular dynamics are 6.7 eV for armchair and 11 eV for zigzag edges.

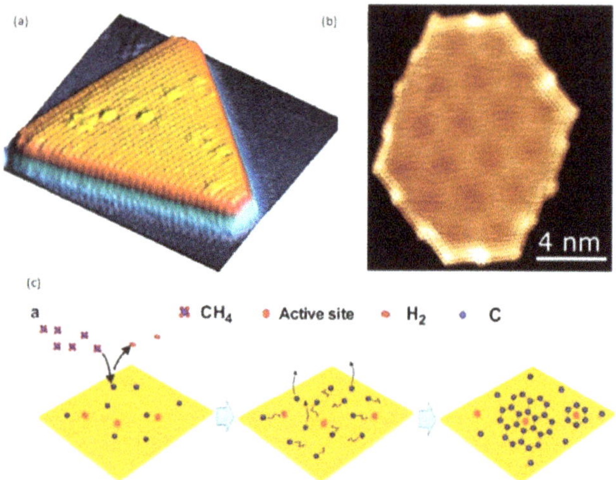

Figure 4.4. (a) Three-dimensional atomic resolution STM of triangular island of graphene on Ni(111). (b) Image of graphene QD on Ir surface. (c) Graphene QDs on Cu surface. Reprinted from [2], Copyright (2023), with permission from Elsevier.

This is to enable the armchair edge to be eliminated in favor of edge of the zigzag edge using electrical current to heat the samples. The edge's dynamics have been studied [30]. The measurements are completed in real time via side spherical aberration-corrected TEM, with specific sensitivity necessary for detecting every atom of carbon that is stable for enough time. The prominent edge structure is zigzag type. Koskinen *et al* [31] have predicted the reconstructed stability edges utilizing the DFT method. The variety of combined stability of higher polygons, heptagons, and pentagons has been noticed [32]. A theoretical investigation has predicted the edge states in Fermi energy vicinity for zigzag edge structures [33]. The mentioned states of the edge have been specified experimentally [28]. Degenerated peak and band have been formed in density of states for graphene ribbons [34]. It is seen utilizing the Hubbard model in an approximation of mean-field in graphene nanoribbons that the electrons occupy edge states to exhibit ferromagnetic case within the edge, and antiferromagnetic case is exhibited between opposite edges of the zigzag [35]. Son *et al* [36] have illustrated, employing DFT, that magnetic properties can be investigated by applying an external electric field across the ribbon. This electric field elevates spin degeneracy while reducing energy bandgap for one spin channel and widening the gap for another. So, we can change the antiferromagnetic coupling between opposite edges into ferromagnetic. There are many investigations of graphene ribbons [37].

The edge's effect has been studied in nanostructured graphene QDs (GQDs). It is clear that the edge types influence the optical properties [38]. In GQDs with zigzag edges, the edge state collapses to degenerate shells on the Fermi level [39]. The relationship between the shell's degeneracy and atom number differences correspond to two sublattices of graphene is obvious [40]. The triangular graphene quantum dot (TGQD) is an example of a degenerate shell system. Therefore, the electronic properties of TGQDs have been studied [41].

4.2.1 Effect of quantization size

The carrier's spatial confinement in nanostructured graphene is anticipated to guide energy spectrum discretization and energy bandgap opening. For graphene ribbons, the bandgap opening has been investigated based on the tight-binding model or resuming of Dirac Hamiltonian [42]. The ribbons and armchair edges oscillate between metallic ground state and insulating as width changes. The energy bandgap size is expected to correlate inversely with the width of the nanoribbon [43]. The experimental works indicate that the energy bandgap opening for the narrowest ribbons, including scaling behavior, agrees with the theoretical investigations [44]. Stampfer *et al* [45] have illustrated GQDs of $D < 100\,\text{nm}$, where effect of quantum confinement plays a crucial role [45]. They have observed peak oscillations of Coulomb blockade as a function of voltage of gate for many peak spacings. The results are in agreement with chaotic Dirac billiards predictions to expect Dirac Fermions behavior confined by an arbitrary shape [46]. The exponential decrease of energy bandgap in terms of Dirac Fermion's diameter has been researched by Recher and Trauzettel [47].

For well-defined edges of a few-nm GDQs, the standing waves of high symmetry are observed using STM [48]. The observations are in agreement with calculations of the DFT and Tmethods. Manninen *et al* [49] have researched the super-shells and shell's structure in the energy spectrum of circular QDs where TGQD is established [49]. Related to their calculations, the TGQD of edge length $\sim$40 nm is required to notice the first super-shell. The Tight-Binding (TB) method predicts energy bandgap opening for arbitrarily shaped GQDs. The exponential decrease of energy bandgap is predicted in terms of the number of atoms [48]. The behavior is different quantitatively for many structures of armchair and zigzag edges, which is connected to the edge states presented in zigzag edges systems [50]. The GQD theory and related properties will be further clarified in the subsequent chapters.

4.2.2 Optical properties of colloidal graphene quantum dots

A solution and colloidal graphene quantum dots (CGQDs) of specific structures has been manufactured [51], while the emission and absorption of CGQDs have been measured [52]. Two types of QDs, first has $N = 168$ and second has $N = 132$ atoms of carbon, are illustrated in figure 4.4.

Mass spectrometry is utilized to determine the number of atoms for each dot, while the symmetry is concluded within infrared vibrational spectra and solution chemistry. The strength and optical response of Coulomb interactions in terms of shape and size, and dielectric constant can be controlled because the CGQDs are suspended in solution. Actually, the spectrum of optical absorption reveals the dependence of absorption edge position on the number of atoms [53]. The phosphorescence and fluorescence spectroscopies [54] demonstrate the existence of an energy bandgap between absorption spectra and emission, interpreted as a function of the energy difference between the triplet and singlet states of the exciton [55].

Problems

1. What is graphene? And, how is it used in the synthesis of QDs?
2. How are graphene QDs synthesized? And, what are some of their unique properties?
3. What are the potential applications of graphene QDs? And, how do they compare to other types of QDs?
4. How can the size and shape of graphene QDs be controlled? And, why is this important for their properties and applications?
5. What are some of the challenges associated with the synthesis and use of graphene QDs? And, how are researchers working to overcome these challenges?

References

[1] Avouris P and Xia F 2012 *MRS Bull.* **37** 1225

[2] Abdelsalam H, Sakr M A S, Saroka V A, Abd-Elkader O H and Zhang Q 2023 *Surf. Interfaces* **40** 103109

[3] Cruz-Silva E, Botello-Mendez A R, Barnett Z M, Jia X, Dresselhaus M S, Terrones H, Terrones M, Sumpter B G and Meunier V 2010 *Phys. Rev. Lett.* **105** 045501

[4] Jia X *et al* 2009 *Science* **323** 1701

[5] Li X, Wang X, Zhang L, Lee S and Dai H 2008 *Science* **319** 1229

[6] Cai J, Ruffieux P, Jaafar R, Bieri M, Braun T, Blankenburg S, Muoth M and Seitsonen A P 2010 *Nature* **466** 470–73

[7] Lu J, Yeo P S E, Gan C K, Wu P and Loh K P 2011 *Nat. Nanotechnol.* **6** 247

[8] Tao C *et al* 2011 *Nat. Phys.* **7** 616

[9] Yazyev O V, Capaz R B and Louie S G 2011 *Phys. Rev.* B **84** 115406

[10] Chico L, Santos H, Ayuela A, Jaskólski W, Pelc M and Brey L 2010 *Acta Phys. Pol.* A **118** 433

[11] Santos H, Ayuela A, Jaskólski W, Pelc M and Chico L 2009 *Phys. Rev.* B **80** 035436

[12] Jaskólski W, Ayuela A, Pelc M, Santos H and Chico L 2011 *Phys. Rev.* B **83** 235424

[13] Müller M, Kbel C and Müllen K 1998 *Chem. Eur. J.* **4** 2099

[14] Tyutyulkov N, Madjarova G, Dietz F and Müllen K 1998 *J. Phys. Chem.* B **102** 10183

[15] Yan X, Li B and Li L-S 2013 *Acc. Chem. Res.* **46** 2254

[16] Mohanty N, Moore D, Xu Z P, Sreeprasad T S, Nagaraja A, Rodriguez A A and Berry V 2012 *Nat. Commun.* **3** 844

[17] Zhang Z, Zhang J, Chen N and Qu L 2012 *Energy Environ. Sci.* **5** 8869

[18] Bunch J S, Yaish Y, Brink M, Bolotin K and McEuen P L 2005 *Nano Lett.* **5** 287

[19] Ponomarenko L A, Schedin F, Katsnelson M I, Yang R, Hill E W, Novoselov K S and Geim A K 2008 *Science* **320** 356

[20] Liu L Z, Tian S B, Long Y Z, Li W X, Yang H F, Li J J and Gu C Z 2014 *Vacuum* **105** 21

[21] Pereira J M Jr, Vasilopoulos P and Peeters F M 2007 *Nano Lett.* **7** 946

[22] Ci L, Xu Z, Wang L, Gao W, Ding F, Kelly K F, Yakobson B I and Ajayan P M 2008 *Nano Res.* **1** 116

[23] Campos L C, Manfrinato V R, Sanchez-Yamagishi J D, Kong J and Jarillo-Herrero P 2009 *Nano Lett.* **9** 2600

[24] Tapazsto L, Dobrik G, Lambin P and Bir L P 2008 *Nat. Nanotechnol.* **3** 397

[25] Hämäläinen S K, Sun Z, Boneschanscher M P, Uppstu A, Ijäs M, Harju A, Vanmaekelbergh D and Liljeroth P 2011 *Phys. Rev. Lett.* **107** 236803
[26] Olle M, Ceballos G, Serrate D and Gambardella P 2012 *Nano Lett.* **12** 4431
[27] Yan Z, Lin J, Peng Z, Sun Z, Zhu Y, Li L, Xiang C, Samuel E L, Kittrell C and Tour J M 2012 *ACS Nano* **6** 9110
[28] Niimi Y, Matsui T, Kambara H, Tagami K, Tsukada M and Fukuyama H 2006 *Phys. Rev.* B **73** 085421
[29] Koo E and Ju S-Y 2015 *Carbon* **86** 318
[30] Neubeck S, You Y M, Ni Z H, Blake P, Shen Z X, Geim A K and Novoselov K S 2010 *Appl. Phys. Lett.* **97** 053110
[31] Koskinen P, Malola S and Häkkinen H 2008 *Phys. Rev. Lett.* **101** 115502
[32] Koskinen P, Malola S and Häkkinen H 2009 *Phys. Rev.* B **80** 073401
[33] Potasz P, Güçlü A D and Hawrylak P 2010 *Phys. Rev.* B **81** 033403
[34] Ezawa M 2006 *Phys. Rev.* B **73** 045432
[35] Wakabayashi K, Sigrist M and Fujita M 1998 *J. Phys. Soc. Jpn.* **67** 2089
[36] Son Y, Cohen M L and Louie S G 2006 *Nature* **444** 347
[37] Wang H and Scarola V W 2012 *Phys. Rev.* B **85** 075438
[38] Guclu A D, Potasz P and Hawrylak P 2010 *Phys. Rev.* B **82** 155445
[39] Yamamoto T, Noguchi T and Watanabe K 2006 *Phys. Rev.* B **74** 121409
[40] Wang W L, Meng S and Kaxiras E 2008 *Nano Lett.* **8** 241
[41] Voznyy O, Güçlü A D, Potasz P and Hawrylak P 2011 *Phys. Rev.* B **83** 165417
[42] Brey L and Fertig H A 2006 *Phys. Rev.* B **73** 235411
[43] Son Y W, Cohen M L and Louie S G 2006 *Phys. Rev. Lett.* **97** 216803
[44] Chen Z, Lin Y-M, Rooks M J and Avouris P 2007 *Physica* E **40** 228
[45] Stampfer C, Güttinger J, Molitor F, Graf D, Ihn T and Ensslin K 2008 *Appl. Phys. Lett.* **92** 012102
[46] Recher P and Trauzettel B 2010 *Nanotechnology* **21** 302001
[47] Subramaniam D *et al* 2010 *Phys. Rev. Lett.* **108** 046801
[48] Ritter K A and Lyding J W 2009 *Nat. Mater.* **8** 235
[49] Manninen M, Heiskanen H P and Akola J 2009 *Eur. Phys. J.* D **52** 143146
[50] Zhang Z Z, Chang K and Peeters F M 2008 *Phys. Rev.* B **77** 235411
[51] Yan X, Cui X and Li L 2010 *J. Am. Chem. Soc.* **132** 5944
[52] Yan X, Cui X, Li B and Li L 2010 *Nano Lett.* **10** 1869
[53] Yan X, Li B, Cui X, Wei Q, Tajima K and Li L 2011 *J. Phys. Chem. Lett.* **2** 1119
[54] Mueller M L, Yan X, McGuire J A and Li L 2010 *Nano Lett.* **10** 2679
[55] Ozfidan I, Korkusinski M, Güçlü A D, McGuire J and Hawrylak P 2014 *Phys. Rev.* B **89** 085310

Chapter 5

Optical studies of Si quantum dots

Optical investigation of Si quantum dots (QDs) reveals unique optical properties resulting from quantum confinement effect. Si QDs exhibit discrete energy levels and enhanced photoluminescence, differing significantly from bulk Si's indirect bandgap characteristics. The key aspects of optical studies include photoluminescence (PL) spectroscopy that analyzes emission spectra of Si QDs when excited by light. The size-dependent emission wavelength shifts, known as quantum size effect, are crucial for tuning optical properties for specific applications.

Advances in the optical characterization of Si QDs contribute to the development of more efficient and versatile nanoscale devices, leveraging the unique properties of quantum-confined silicon for next-generation technologies.

5.1 Introduction

QD semiconductors have attracted a lot of interest from engineers and scientists of different disciplines because of their unique properties and flexible processibility [1]. Luminescent nanocrystals are desirable for optoelectronics such as LEDs [2]. Because of the large surface-to-volume ratio of nanoparticles, the main reason for weak efficiency of luminescence is attributed to non-radiative recombination for charge carriers of light-generated at surface-traps [3]. QDs can be employed in solar panels to improve efficiency, and for green and clean applications. The dielectric functions, photoemission spectra, optical constants, and bandgaps depend on the bonding and chemical properties, alloy properties, frequency, and wavevector, which have been estimated for different materials with the pseudo-potential method [4].

Different QD solar concentrators have been researched [5]. Kennedy *et al* [6] investigated the lower optical efficiency of single-plate quantum dot solar concentrators (QDSCs) due to their low luminescent quantum yields, large overlap between QD emission, and absorption spectra of present commercially-available visible-emitting QDs. In addition, they found that the reabsorption of QD emitted photons could be

doi:10.1088/978-0-7503-5704-3ch5

decreased by near-infrared emitting QDs, thus reducing the loss of the escape cone improves the optical concentration ratio and efficiency. It is clear that the mentioned loss via Monte-Carlo modeling accounts of ~57% for incident photons that are absorbed by visible-emitting QDSCs. Meanwhile, Gallagher *et al* [7] have depicted new non-tracked concentrators that utilize QD technology to make a practical suggestion for a fluorescent dye-solar concentrator. They have indicated that QDSCs can be incorporated into the façades of buildings as photovoltaics (PVs) to convert solar energy into electrical current, and as a result have manufactured small QDSCs. Gallagher *et al* [8] achieved transparent host materials and spectroscopic measurements for different diameters of QDs, and presented evidence of high efficiency of QDs that are necessary for QDSCs. They determined the optimal material for absorption characteristic-based QDSCs and the optimum commercial model of QDs has been selected via steady-state photoluminescence excitation spectroscopy, photoluminescence, and absorption of QDs in matrices of solid and solution.

Udipi *et al* [9] submitted a simulation work for electron density distribution and potential energy profile in 200 nm Si QDs. For the equation of continuity solution, the approximation of efficient difference suggested by Scharfetter and Gummel [10] has been expanded to three dimensions. Essentially, they followed a two-dimensional approach because Selberherr *et al* [11] have extended two dimensions to three dimensions. Meanwhile, Zhong and Liu [12] have used the electron Raman scattering (ERS) process associated with bulk-like interface optical (IO) phonon and longitudinal optical (LO) modes of $GaAs/Ga_{1-x}Al_xAs$ QDs, in addition to their efforts to select rules of singularities and processes of Raman spectra of different concentrations (x). Hasaneen *et al* [13] developed an electrical characteristic simulating model for non-volatile QD floating-gate memory. They calculated the rate of tunneling electrons by Bardeen's transition-Hamiltonian, followed by potential energies and wavefunction of the quantum well (QW) channel and QD gate via Poisson and Schrödinger equations for a self-consistent numerical solution. They found that by modifying the QD charge, the values of a resistor can be modified by 40%. From the experimental viewpoint, Schmidt *et al* [14] synthesized buried QW of InGaAs/GaAs single QWs utilizing selective wet etching of the top barrier and high-resolution electron beam lithography. They produced QDs of diameters below 20 nm to maintain high luminescence efficiency for small sizes. They also noticed a blue shift of 9 meV for QDs size, 23 nm.

The novel material investigations have attracted much interest, especially to research the diameter dependence of QDs in connection with the nearest atoms bonds. To control the diameter of QDs, it is important to explore the QDs' potential with their diameter. In this situation, we have followed the full potential-linearized augmented plane wave (FP-LAPW) procedure to test our model's validity [15] for dot diameter-based QD potential less than 52 nm, including an investigation of Si's optical properties.

5.2 Technique of Si quantum dots

Chatten *et al* [16] have presented a QD concentrator, theoretically and experimentally. First, to develop a self-consistent thermodynamic model using FP-LAPW for planar concentrators to explore the three-dimensional flux model showing an excellent agreement with experimental work, it will be advantageous to employ ab initio methods within FP-LAPW to study the electronic properties. Al-Douri *et al* [17–19] developed an accurate scheme for the band structure calculation method. In the FP-LAPW method, the unit cell is divided into an interstitial region and non-overlapping muffin-tin spheres around the sites of atoms. Among both types for different basis sets, the Kohn-Sham equation is based on density functional theory [20] to be solved in a self-consistent scheme. The potential of exchange correlation is treated utilizing generalized gradient approximation [21] where orbital of Si ($3s^2 3p^2$) is used as valence band for total energy calculations. Furthermore, Engel and Vosko's generalized gradient approximation (EVGGA) formalism [22] is employed to study the optical and electronic properties.

5.3 Investigations of Si quantum dots

Covalent semiconductors have four-fold coordination, which is attributed to their low density and because their nearest neighbors are bounded via overlapping hybridized orbitals that are known as hybrid of sp^3 with tetrahedral direction. Therefore, it is possible to tune the bandgap utilizing dot diameter. Table 5.1 presents the computed direct ($\Gamma \rightarrow \Gamma$) and indirect ($\Gamma \rightarrow X$) and ($\Gamma \rightarrow L$) bandgap values utilizing EVGGA for Si of different diameters in comparison with experimental [23] and theoretical data [24]. Our computed ($\Gamma \rightarrow X$) bandgaps are overestimated in comparison with the available results utilizing EVGGA.

The Si element has a narrow bandgap. Due to using in infrared light detection and generation, the variation of bandgap for dot diameters represents an attractive study. As presented in table 5.1, the bandgap values decrease at point X. From figure 5.1, it is noticed that bandgap values vary with dot diameter, showing enlarging along $\Gamma \rightarrow X$ and reducing along $\Gamma \rightarrow L$ and $\Gamma \rightarrow \Gamma$ to confirm that Si has an indirect bandgap.

Table 5.1. Calculated principal energy gaps for Si (in eV) at different diameters (in nm) in comparison with others.

Dot diameter	E_g ($\Gamma-\Gamma$)	E_g ($\Gamma-X$)	E_g ($\Gamma-L$)
54.3	2.742	1.436; 1.11[a,b]	2.028
54	2.747	1.396	2.094
53.6	2.751	1.352	2.164
53.3	2.757	1.272	2.279
53	2.752	1.345	2.174
52.7	2.759	1.233	2.332

[a] Reference [20] experiment.
[b] Reference [21].

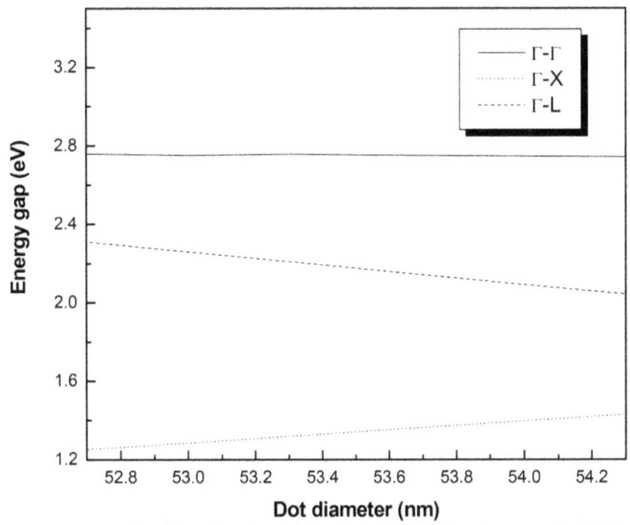

Figure 5.1. Dot diameter dependence of calculated direct (Γ→Γ) and indirect (Γ→X) and (Γ→L) energy gaps of Si.

Table 5.2. Calculated QDs potential for Si (in mV) at different diameters (in nm) in comparison with others.

Dot diameter	P_{QD} Cal.	P_{QD} [9]
54.3	1.051	$\leqslant 1$
54	1.021	
53.6	0.989	
53.3	0.931	
53	0.984	
52.7	0.902	

The gap between the conduction band minimum at point X and valence band maximum at point Γ is called the bandgap and is calculated using FP-LAPW. By applying our model [15], the QD potential has been estimated by the formula:

$$P_{QD} = \frac{b}{a}.\ E_{g\Gamma X}.\ 10^{-3}.\ \lambda \tag{5.1}$$

where constant refers to $\frac{b}{a}$ in eV^{-1} (see table 5.4 in reference [15]) and parameter refers to λ for II–VI ($\lambda = 2$), III–V ($\lambda = 4$) and group-IV ($\lambda = 6$) semiconductors in V.

When the diameter decreases, the strong sp^3 covalent bonding affects and characterizes the covalent structure. The diameter dependence of QDs is a result of the difference in the QDs' potential. Table 5.2 displays the calculated diameter dependence of QDs. The separation between increase and decrease of the QDs'

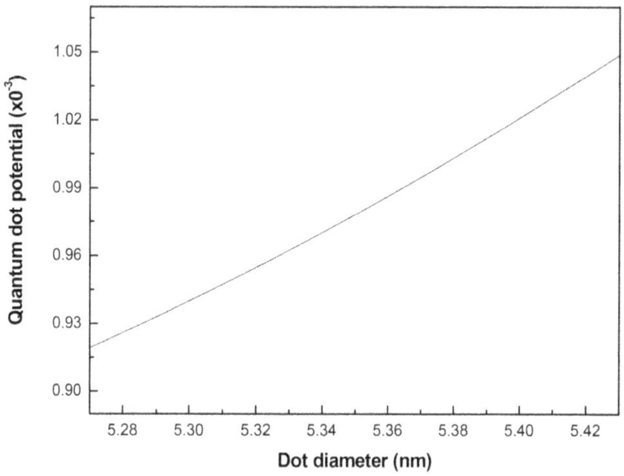

Figure 5.2. QDs potential as a function of dot diameter for Si.

potential is called the critical dot diameter, which correlates with transition pressure (P_t) that is necessary to compute molar free energies difference. The difference of Gibbs free-energy (ΔG_t) between different compounds having tetrahedral coordination is given by $P_t\,\Delta V/V_t$ (in kJ mol^{-1}). The energy values are large for small bond lengths. The decreasing dot diameter-based QDs' potential confirms the position change of energy bands at principal points; $\Gamma \rightarrow \Gamma$, $\Gamma \rightarrow X$ and $\Gamma \rightarrow L$, as presented in table 5.1.

There is a direct correlation between the QDs' potential and dot diameter until a critical value is reached, as analyzed by a random behavior (table 5.2) and affirmed by figure 5.2. As a result, the fluctuation of the QDs' potential appears. Our calculated QD values are in agreement with other data [9]. So, the variation in the QDs' potential refers to tunneling electron control across QDs.

The microscopic atomic interactions represented by refractive index n are an attractive optical parameter. Theoretically, there are two approaches: the local polarizability of these entities and the refractive index will be related to the density [25]. Therefore, different attempts have been made to correlate the energy gap E_g and refractive index within specific relationships [4]. The mentioned relations of n are independent of incident photon energy and temperature. So, a review is done for different relations of E_g and n. A linear for n as a function of E_g is given by Ravindra *et al* [26]:

$$n = \alpha + \beta E_g \tag{5.2}$$

where $\alpha = 4.048$ and $\beta = -0.62\,\text{eV}^{-1}$. Light refraction and dispersion will be inspired. Herve and Vandamme [27] proposed an empirical relation as follows:

$$n = \sqrt{1 + \left(\frac{A}{E_g + B}\right)^2} \tag{5.3}$$

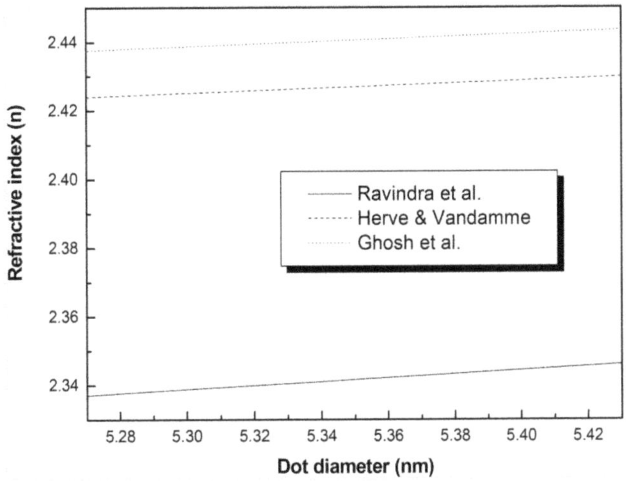

Figure 5.3. Dot diameter dependence of the refractive index (n) for Si.

Table 5.3. Calculated dot diameter-based refractive indices of Si utilizing Ravindra *et al* [26], Herve and Vandamme [27], and Ghosh *et al* [28] models corresponding to optical dielectric constant.

Dot diameter	n	ε_∞
54.3	2.347[a] 2.4297[b] 2.4441[c] 3.882[d]	5.5084[a] 5.9034[b] 5.9736[c] 11.68[e]
54	2.344[a] 2.4296[b] 2.4422[c]	5.4943[a] 5.9029[b] 5.9643[c]
53.6	2.342[a] 2.4266[b] 2.4406[c]	5.484[a] 5.8883[b] 5.9565[c]
53.3	2.338[a] 2.4246[b] 2.4383[c]	5.466[a] 5.8786[b] 5.9453[c]
53	2.341[a] 2.4263[b] 2.4402[c]	5.480[a] 5.886[b] 5.9545[c]
52.7	2.337[a] 2.4240[b] 2.4335[c]	5.461[a] 5.8757[b] 5.9219[c]

[a] Reference [26].
[b] Reference [27].
[c] Reference [28].
[d] References [32, 33] experiment.
[e] Reference [34] experiment.

where $A = 13.6$ eV and $B = 3.4$ eV. For group-IV semiconductors, Ghosh *et al* [28] have published an empirical relationship based on the band structure and quantum dielectric considerations of Penn [29] and Van Vechten [30]:

$$n^2 - 1 = \frac{A}{(E_g + B)^2} \tag{5.4}$$

where the appropriate average energy gaps are $A = 8.2E_g + 134$, $B = 0.225E_g + 2.25$ and $(E_g + B)$. So, employing the three mentioned models of refractive index, the variation of the dot diameter-based n has been calculated. Figure 5.3 illustrates the results and table 5.3 displays the calculated n values. This is verified by calculating optical dielectric constant ε_∞ that is dependent on the refractive index. Where $\varepsilon_\infty = n^2$ [31]. It is obvious that n calculated utilizing Ghosh *et al* model [28] is in accordance with experimental results, where the reflectivity is necessary for solar cells. The linear dependence of dot diameter of Si is noticed and the refractive index for small diameter tends to shift towards blue–green, which means that low reflection and the high absorption spectrum are attributed to the increasing efficiency of the solar cells.

The FP-LAPW has proved to be a good method to calculate electronic properties to investigate optical studies of lower reflectivity for IV-compounds to present evidence suitability of dot diameter 53.6 nm for optoelectronic applications as a novel trend for alternative materials for QD techniques.

Problems

1. What are Si QDs? And, what makes them interesting for optical studies?
2. How are Si QDs synthesized? And, what are some of the challenges associated with their synthesis?
3. What are some of the optical properties of Si QDs? And, how can they be investigated?
4. What are some of the potential applications of Si QDs in optoelectronics? And, how do their optical properties impact these applications?
5. What are some of the current research areas and future directions in the investigation of optical studies of Si QDs?

References

[1] Landes C, Burda C, Braun M and El-Sayed M A 2001 *J. Phys. Chem.* B **105** 2981

[2] Gerion D, Pinaud F, Williams S C, Parak W J, Zanchet D, Weiss S and Alivisatos A P 2001 *J. Phys. Chem.* B **105** 8861

[3] Sondi I, Siiman O and Matijevic E 2004 *J. Colloid Interface Sci.* **275** 503

[4] Al-Douri Y, Feng Y P and Huan A C H 2008 *Solid State Commun.* **148** 521

[5] Al-Douri Y 2004 *Mater. Chem. Phys.* **88** 339

[6] Kennedy M, McCormack S J, Doran J and Norton B 2009 *Sol. Energy* **83** 978

[7] Gallagher S J, Norton B and Eames P C 2007 *Sol. Energy* **81** 813

[8] Gallagher S J, Rowan B C, Doran J and Norton B 2007 *Sol. Energy* **81** 540

[9] Udipi S, Vasileska D and Ferry D K 1996 *Superlattices Microstruct.* **20** 343

[10] Scharfetter D L, Gummel H K and Trans I E E E 1969 *Electron. Devices* **ED-16** 64

[11] Selberherr S, Shutz A, Potzl H W and Trans I E E E 1980 *Electron. Devices* **ED-27** 1540

[12] Zhong Q-H and Liu C-H 2009 *Solid-State Electron* **53** 134

[13] Hasaneen E, Heller E, Bansal R, Huang W and Jain F 2004 *Solid-State Electron* **48** 2055

[14] Schmidt A, Forchel A, Faller F, Itskevich I E and Vasiliev A 1994 *Solid-State Electron* **37** 1101

[15] Al-Douri Y 2009 Quantum dot modeling of semiconductors ed A H Reshak *Advances in Condensed Matter Physics* (Kerala, India: Research Signpost)

[16] Chatten A J, Barnham K W J, Buxton B F, Ekins-Daukes N J and Malik M A 2004 *Semiconductors* **38** 909

[17] Al-Douri Y, Hussain Reshak A, Baaziz H, Charifi Z, Khenata R, Ahmad S and Hashim U 2010 *Sol. Energy* **84** 1979

[18] Al-Douri Y 2004 *Solid State Commun.* **132** 465
[19] Blaha P, Schwarz K, Madsen G K H, Kvasnicka D and Luitz J 2001 *WIEN2K* (Austria: Technische Universitat, Wien)
[20] Kohn W and Sham L J 1965 *Phys. Rev.* **140** A1133
[21] Perdew J P, Burke S and Ernzerhof M 1996 *Phys. Rev. Lett.* **77** 3865
[22] Engel E and Vosko S H 1993 *Phys. Rev.* B **47** 13164
[23] Tsidilkovski I M 1982 *Band Structure of Semiconductors* (Oxford: Pergamon)
[24] Humphreys R G, Rossler V and Cardona M 1978 *Phys. Rev.* B **18** 5590
[25] Balzaretti N M and da Jornada J A H 1996 *Solid State Commun.* **99** 943
[26] Ravindra N M, Auluck S and Srivastava V K 1979 *Phys. Status Solidi* B **93** K155
[27] Herve P J L and Vandamme L K J 1995 *J. Appl. Phys.* **77** 5476
[28] Ghosh D K, Samanta L K and Bhar G C 1984 *Infrared Phys.* **24** 34
[29] Penn D R 1962 *Phys. Rev.* **128** 2093
[30] Van Vechten J A 1969 *Phys. Rev.* **182** 891
[31] Samara G A 1983 *Phys. Rev.* B **27** 3494
[32] Aspnes D E and Theeten J B 1980 *J. Electrochem. Soc.* **127** 1359
[33] Palik E D 1985 *Handbook of Optical Constants of Solids* (Boston, MA: Academic)
[34] Gray P R, Hurst P J, Lewis S H and Meyer R G 2009 *Analysis and Design of Analogue Integrated Circuits* (New York: Wiley)

Chapter 6

Further optical properties of CdX (X = S, Te) compounds under quantum dot diameter effect: *ab initio* method

This chapter explores the optical characteristics of cadmium chalcogenide (CdS and CdTe) quantum dots (QDs) as a function of their size using first-principles calculations. The quantum confinement effects in these semiconductor nanocrystals significantly alter their electronic and optical properties compared to their bulk counterparts. The key points of this study include quantum confinement, which focuses on how reducing the QD diameter influences the energy levels and bandgap of CdS and CdTe QDs. Smaller diameters lead to larger bandgaps and shifts in absorption and emission spectra due to quantum confinement effects. Meanwhile, the study of optical absorption and emission uses *ab initio* methods to calculate the size-dependent absorption and emission spectra of CdS and CdTe QDs.

6.1 Introduction

The large energy gap, green–blue optoelectronics, and use in solar cells are the main features of II–VI materials [1]. Theoretical band structure calculation methods for binary materials have been employed to study the electronic properties and different models have been used [2]. A comparison of different theoretical and experimental results is used to test different approaches. The transition from the coordinated number $N_c = 4–6$ is well demonstrated by the use of a computational method based on total energy calculations [3].

Miguez *et al* [4] investigated optoelectronic properties of shallow acceptors and donors of GaAs-$Ga_{1-x}Al_x$ QDs under combined effect of intense laser and isotropic hydrostatic pressure. They computed the optoelectronic properties as a function of laser field amplitude, dot size, and hydrostatic pressure. They have found a direct correlation between impurity binding energy and pressure, and an inverse correlation between impurity binding energy and laser field amplitude, while the pressure

doi:10.1088/978-0-7503-5704-3ch6

effect is more noticed for donor than for acceptor impurities, especially for small QDs. The structure of $In_{0.5}Ga_{0.5}As$ stacked QDs is analyzed and characterized utilizing photoluminescence (PL), high-resolution x-ray diffraction (HR-XRD) and atomic force microscopy [5]. The authors investigated stacked structured dot size and density to prove the strong influence by dot formation under-structure and spacer layers. The formation of dots is not vertically aligned since HR-XRD analysis has given different peaks on n-stacked QDs structure, additionally it is attributed to dot size, density, and composition in the stacked structures. Meanwhile, Panchal *et al* [6] have researched Si QDs embedded into silicon nitride (SiNx) of different characteristics and fabrication techniques. They elaborated the SiNx advantages in comparison with silicon dioxide (SiO_2) for Si-QDs from the electronics viewpoint to outline different techniques of fabrication of different conditions of deposition. Udipi *et al* [7] simulated electron density distribution and potential energy profile of 200 nm Si QDs. For the continuity equation of the solution, the efficient approximations of difference suggested by Scharfetter and Gummel [8] have been extended to three dimensions. Essentially, the authors followed up the two dimension approach of Selberherr *et al* [9] and extended it to three dimensions.

Liu *et al* [10] synthesized center hollow TiO_2 and ZnO nanotubes arrays via sol-gel and chemical etching techniques, respectively. Moreover, for TiO_2 and ZnO nanotubes applications, they successfully fabricated and characterized cell performance and dye-sensitized solar cells (DSSCs), respectively. They obtained efficient ZnO-based DSSCs of 1.2% and TiO_2-based DSSCs of 2.1% nanotubes. Das and Sokol [11] synthesized nanostructured ZnO with the solution method at low temperature. They characterized the structure, distribution, and size of ZnO utilizing neutron scattering technique and XRD, in addition to fabricating bulk heterojunction solar cells of hybrid polymer-metal oxide via blending regioregular poly(3-hexylthiophene) (P3HT) and ZnO through flow coating and solution process on a specific flexible substrate. They found that when PL decreases for more than 79% of ZnO:P3HT composites, it is an indication of high efficiency. The solar cells of I_{sc} and V_{oc} have 6.5 mA cm^{-2} and 0.33 V, respectively. The stability and performance of solar cells have been researched by UV-vis. Meanwhile, Badescu and Badescu [12] have improved the efficiency of solar cells by up-converting the sub-energy-gaps by adding an up-converter that is able to convert incident solar photons of low-energy into high energy photons. Their novelty is that they take into consideration the solar cell's refractive index, including the converter. They concluded that: (1) the efficiency of solar energy increases for cell and rear converter (C–RC) system in comparison with solar a cell operating alone, especially at high concentration ratios; (2) the efficiency of solar energy for C–RC system increases as up-converter refractive indices and solar cells increase; and (3) the efficiency of solar energy does not undergo an increase when adding a front up-converter to solar cells, regardless of what the concentration ratio is.

This research is especially attractive when trying to obtain further information of the diameter dependence of materials [13]. It is able to correlate bonds of nearest atoms and diameter dependence. By managing diameter dependence evaluation, it is possible to connect the QDs' diameter and potential. So, it is employed to verify the validity of our QD potential model [14]. It is aimed to calculate dot diameter-based QDs' potential for less than 65 and 55 nm for CdTe and CdS, respectively, by full potential linearized augmented plane wave (FP-LAPW) method to research the optical properties.

6.2 Technique of CdX (X = S, Te) quantum dots

This research is carried out utilizing the FP-LAPW method as implemented in WIEN2K code [15]. It is known that the unit cell is divided into an interstitial region and non-overlapping muffin-tin spheres around sites of atoms. Among them, different basis sets are utilized, so the Kohn–Sham equation based on DFT [16] is solved in a self-consistent scheme. The potential of exchange correlation is treated via generalized gradient approximation (GGA) [17] for orbitals of Te ($4d^{10}5s^25p^4$), S ($3s^23p^4$) and Cd ($4d^{10}5s^2$) are addressed as valence electrons for total energy calculations. Furthermore, the formalism of Engel and Vosko's generalized gradient approximation (EVGGA) [18] is utilized to calculate the optoelectronic properties.

6.3 Investigation of CdX (X = S, Te) quantum dots

Four-fold coordination is a feature of covalent semiconductors. Where the density is low and the nearest neighbors are overlapped by hybridized orbitals, they are known as sp^3 hybrids with tetrahedral direction. So, it is possible to tune the bandgap utilizing dot diameter. The calculated energy gap values of direct (Γ→Γ) and indirect (Γ→L) and (Γ→X) using EVGGA of dot diameter-based CdTe and CdS are given in table 6.1,

Table 6.1. Calculated energy gaps for dot diameter-based CdS and CdTe in eV in nm compared to experimental and theoretical data.

QD's diameter	E_g (Γ−Γ)	E_g (Γ−X)	E_g (Γ−L)
Bulk	2.359, 2.42[a] 2.361[b]	***CdS***	3.432
55.75	2.352	4.626	3.261
56.17	2.292	4.584	3.163
56.37	2.262	4.432	3.114
57.78	2.051	4.358	2.788
58.17	1.991	3.871	2.699
58.56	1.993	3.741	2.613
Bulk	1.368 1.8[c] 1.8[d]	3.619	2.288
66.36	1.133	***CdTe***	2.267
66.17	1.163	3.241	2.244
66.02	1.1639	3.05	2.229
65.86	1.208	3.067	2.244
65.59	1.253	3.081	2.2596
65.33	1.283	3.0819	2.2745
		3.096	
		3.111	

[a] Reference [19] experiment.
[b] Reference [21] theory.
[c] Reference [20] experiment.
[d] Reference [22] theory.

and compared with experimental results [19, 20] and theoretical data [21, 22]. Our values of ($\Gamma \rightarrow \Gamma$) bandgap are underestimated in comparison with the experimental results, which is attributed to the use of EVGGA. Therefore, CdTe and CdS can be regarded as direct energy gap semiconductors. Due to using in infrared light detection and generation, the variation of dot diameter-based energy gaps is studied. Table 6.1 presents energy gaps that are correlated inversely with the dot diameter, as verified by figure 6.1.

The energy gaps between conduction band minimum (CBM) at point X and valence band maximum (VBM) at point Γ are calculated by the FP-LAPW method. Using our model [14], the QDs' potential has estimated:

$$P_{\mathrm{QD}} = \frac{b}{a}.\ E_{g\Gamma X}.\ 10^{-3}.\ \lambda \tag{6.1}$$

where $\frac{b}{a}$ is constant (in eV^{-1}) (see table 4 in reference [14]) and parameter refers to λ for II–VI ($\lambda = 2$), III–V ($\lambda = 4$) and group-IV ($\lambda = 6$) (in V). The correlation between pressure effect changes and QD's diameter is defined. If the QDs' diameter changes, the sp^3 covalent bond is affected. This diameter dependence is an immediate result of QDs potential difference. The calculated dot diameter-based QD potential is presented in table 6.2.

The separation of decreasing and increasing of QD potential is called the critical dot diameter. The correlation between diameter dependence and transition pressure (P_t) is necessary to calculate from molar free energies difference. The diameter-based Gibbs free-energy difference ΔG_t is given by $P_t\ \Delta V/V_t$ (in kJ mol^{-1}). It seems that energies are large for small bond length. Changes in the dot diameter-based QD potential are verified by a change in the energy gaps, as presented in table 6.1. The inverse correlation between QD's potential and dot diameter is presented in table 6.2 and verified by figure 6.2. The linear relationship of CdS and nonlinear of CdTe is shown. As a result, QD potential fluctuation appears. The calculated values of QD potential are in agreement with previous work [7]. So, the QD potential variation indicates control of electron tunneling across QDs.

The optical properties represented by refractive index n are necessary for atomic interactions. So, there are two different approaches: local polarizability and refractive index [23]. Therefore, different simple relationships have been subjected to correlate n and E_g [24]. The mentioned relations of n are independent of incident photon energy and temperature. Various relations of n and E_g will be studied. Ravindra *et al* [25] have submitted a linear form of n in terms of E_g:

$$n = \alpha + \beta E_g \tag{6.2}$$

where $\alpha = 4.048$ and $\beta = -0.62\ \mathrm{eV}^{-1}$. Light refraction and dispersion will be inspired. Herve and Vandamme [26] proposed an empirical relation as follows:

$$n = \sqrt{1 + \left(\frac{A}{E_g + B}\right)^2} \tag{6.3}$$

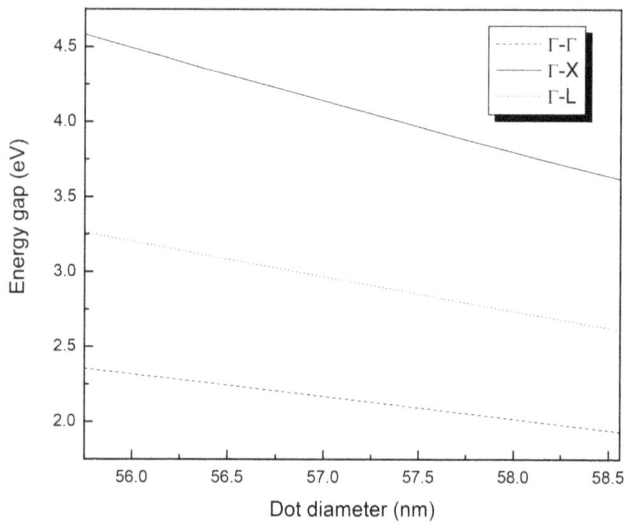

(a)

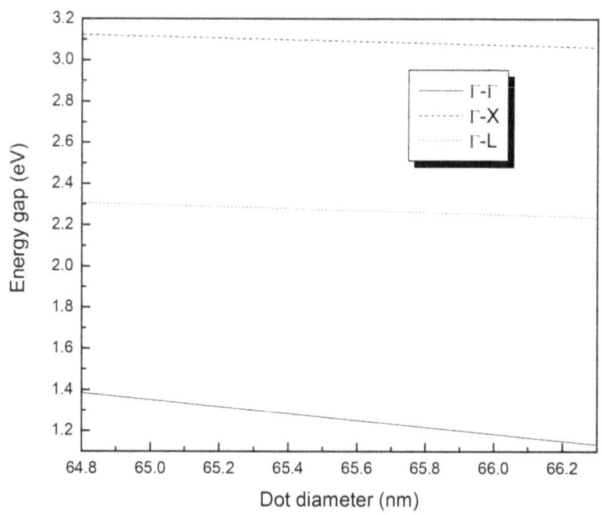

(b)

Figure 6.1. Calculated dot diameter-based energy gaps of direct (Γ → Γ) and indirect (Γ → X) and (Γ → L) of CdS (a), CdTe (b).

where $A = 13.6\,\text{eV}$ and $B = 3.4\,\text{eV}$. For group II–IV semiconductors, Ghosh *et al* [27] published an empirical relationship based on the band structure and quantum dielectric considerations of Penn [28] and Van Vechten [29]:

Table 6.2. Calculated dot diameter-based QD potential of CdS and CdTe in mV compared to other data.

QD's diameter	P_{QD} cal.	P_{QD} [7]
55.75	***CdS***	1
56.17	0.874	$\leqslant 1$
56.37	0.845	
57.78	0.831	
58.17	0.738	
58.56	0.713	
66.36	0.690	
66.17	***CdTe***	
66.02	0.582	
65.86	0.585	
65.59	0.5885	
65.33	0.588	
	0.590	
	0.5937	

$$n^2 - 1 = \frac{A}{(E_g + B)^2} \tag{6.4}$$

where $A = 8.2E_g + 134$, $B = 0.225E_g + 2.25$ and $(E_g + B)$ refers to an appropriate average energy gap of the material. Thus, using these three models, the variation of n with dot diameter has been calculated. The results are displayed in figure 6.3. The calculated refractive indices and the dielectric optical constants of the end-point compounds are investigated and listed in table 6.3.

This is verified by the calculation of the optical dielectric constant ε_∞, which depends on the refractive index. Note that $\varepsilon_\infty = n^2$ [30]. It is known that n utilizing Herve and Vandamme model [26] is in agreement with experimental data because reflectivity is required to enhance the efficiency of solar cells. In addition, dot diameter-based CdS and CdTe optoelectronic properties are noticed, where n of small diameter shifts to blue–green, which means that low reflection and high absorption are attributed to increased efficiency of solar cells.

The FP-LAPW method has proved to be a good technique for calculating electronic properties, its validity has been confirmed, the optical properties of low reflectivity value of II–IV materials have been investigated, and it has been proven that 65.33 and 55.75 nm dot diameters for CdTe and CdS, respectively, are appropriate for photovoltaic applications, and for new studies and realizations of QDs.

Problems

1. What are CdX (X = S, Te) compounds? And, what makes them interesting to study their optical properties using *ab initio* methods?

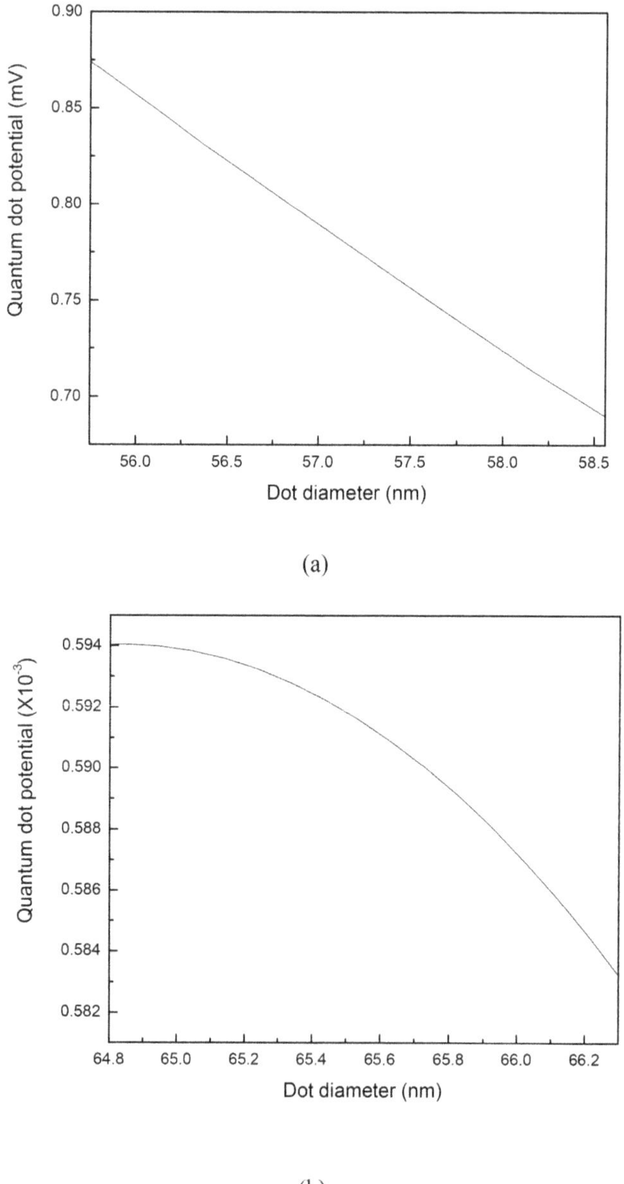

Figure 6.2. Dot diameter-based QD potential of CdS (a) and CdTe (b).

2. How do the optical properties of CdX compounds change as the diameter of the QDs is varied? And, how can this be studied using *ab initio* methods?
3. What are some of the challenges associated with modeling the optical properties of CdX compounds using *ab initio* methods? And, how can these be overcome?

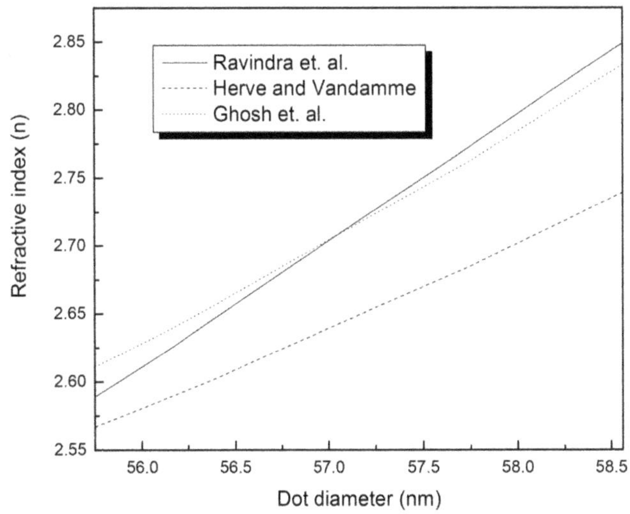

(a)

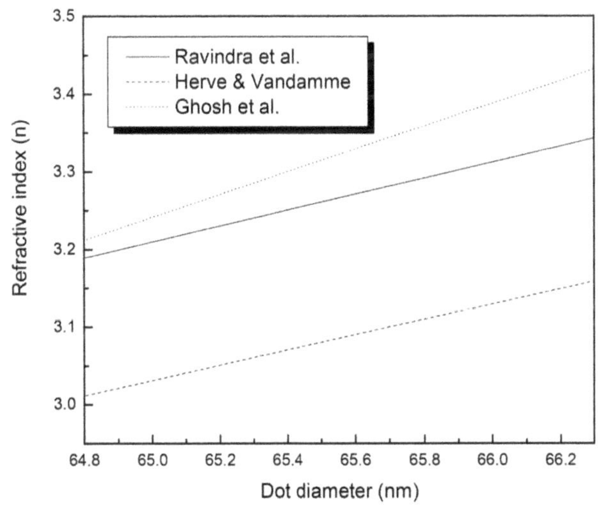

(b)

Figure 6.3. Dot diameter-based refractive index of CdS (a) and CdTe (b).

4. What are some of the potential applications of CdX QDs in optoelectronics? And, how do their optical properties impact these applications?
5. What are some of the current research areas and future directions in the study of optical properties of CdX QDs using *ab initio* methods?

Table 6.3. Calculated dot diameter-based refractive indices of CdS and CdTe using Ravindra *et al*'s [25], Herve and Vandamme's [26], and Ghosh *et al*'s [27] models corresponding to optical dielectric constant.

QD's diameter	n	ε_∞
55.75	***CdS***	6.702[a] 2.589[b] 2.817[c]
56.17	2.589[a] 2.567[b] 2.611[c] 2.38[d]	6.895[a] 6.708[b] 6.969[c]
56.37	2.626[a] 2.590[b] 2.640[c]	6.996[a] 6.765[b] 7.049[c]
57.78	2.645[a] 2.601[b] 2.655[c]	7.706[a] 7.360[b] 7.645[c]
58.17	2.776[a] 2.687[b] 2.765[c]	7.912[a] 7.360[b] 7.834[c]
58.56	2.813[a] 2.713[b] 2.799[c]	8.116[a] 7.503[b] 8.025[c]
66.36	2.849[a] 2.739[b] 2.833[c]	11.189[a] 9.998[b] 11.826[c]
66.17	***CdTe***	11.068[a] 9.882[b] 11.629[c]
66.02	3.345[a] 3.1627[b] 3.439[c] 2.7[d]	11.064[a] 9.879[b] 11.623[c]
65.86	3.3269[a] 3.1437[b] 3.4102[c]	10.883[a] 9.709[b] 11.343[c]
65.59	3.3263[a] 3.1432[b] 3.4093[c]	10.699[a] 9.541[b] 11.068[c]
65.33	3.299[a] 3.116[b] 3.368[c]	10.575[a] 9.431[b] 10.89[c]
	3.271[a] 3.089[b] 3.327[c]	
	3.252[a] 3.071[b] 3.300[c]	

[a] Reference [25].
[b] Reference [26].
[c] Reference [27].
[d] Reference [25] experiment.

References

[1] Hasse M A, Qiu J, DePuydt J M and Cheng H 1991 *Appl. Phys. Lett.* **59** 1272
[2] Dornhaus R, Nimtz G and Richter W 1976 *Solid State Physics* (Berlin: Springer)
[3] Chelikowsky J R 1987 *Phys. Rev.* B **35** 1174
[4] Miguez A, Franco R and Silva-Valencia J 2010 *Int. J. Mod. Phys.* B **24** 5761
[5] Aryanto D, Othaman Z, Ameruddin A S and Ismail A K 2010 *Nano* **5** 127
[6] Panchal A K, Rai D K, Mathew M and Solanki C S 2009 *Nano* **4** 265
[7] Udipi S, Vasileska D and Ferry D K 1996 *Superlattices Microstruct.* **20** 343
[8] Scharfetter D L and Gummel H K 1969 *IEEE Trans. Electron Devices* **ED-16** 64
[9] Selberherr S, Shutz A and Potzl H W 1980 *IEEE Trans. Electron Devices* **ED-27** 1540
[10] Liu Z, Liu C, Ya J and Lei E 2011 *Renew. Energy* **36** 1177
[11] Das N C and Sokol P E 2010 *Renew. Energy* **35** 2683
[12] Badescu V and Badescu A M 2009 *Renew. Energy* **34** 1538
[13] Al-Douri Y, Khenata R and Reshak A H 2011 *Sol. Energy* **85** 2283
[14] Al-Douri Y 2009 Quantum dot modeling of semiconductors in: A H Reshak ed *Advances in Condensed Matter Physics* (Kerala, India: Research Signpost)
[15] Blaha P, Schwarz K, Madsen G K H, Kvasnicka D and Luitz J 2001 *WIEN2K* (Wien, Austria: Technische Universitat)
[16] Kohn W and Sham L J 1965 *Phys. Rev.* **140** A1133
[17] Perdew J P, Burke S and Ernzerhof M 1996 *Phys. Rev. Lett.* **77** 3865
[18] Engel E and Vosko S H 1993 *Phys. Rev.* B **47** 13164

[19] Lincot D and Hodes G 2006 *Chemical Solution Deposition of Semiconducting and Non-Metallic Films* (Pennington, NJ: Electrochemical Society)
[20] Tsidilkovski I M 1982 *Band Structure of Semiconductors* (Oxford: Pergamon)
[21] Feng Y P, Teo K L, Li M F, Poon H C, Ong C K and Xia J B 1993 *J. Appl. Phys.* **74** 3948
[22] Al-Douri Y, Abid H, Zaoui A and Aourag H 2001 *Physica* B **301** 295
[23] Balzaretti N M and da Jornada J A H 1996 *Solid State Commun.* **99** 943
[24] Al-Douri Y, Feng Y P and Huan A C H 2008 *Solid State Commun.* **148** 521
[25] Ravindra N M, Auluck S and Srivastava V K 1979 *Phys. Status Solidi* B **93** K155
[26] Herve P J L and Vandamme L K J 1995 *J. Appl. Phys.* **77** 5476
[27] Ghosh D K, Samanta L K and Bhar G C 1984 *Infrared Phys.* **24** 34
[28] Penn D R 1962 *Phys. Rev.* **128** 2093
[29] Van Vechten J A 1969 *Phys. Rev.* **182** 891
[30] Samara G A 1983 *Phys. Rev.* B **27** 3494

Chapter 7

Ab initio method for optical investigations of $CdS_{1-x}Te_x$ alloys under quantum dot diameter effect

This chapter examines the optical properties of cadmium sulfide-telluride ($CdS_{1-x}Te_x$) alloy quantum dots (QDs) as influenced by their size using first-principles calculations. This research delves into how the quantum confinement effects impact the optical characteristics of these semiconductor nanocrystals, which are important for various optoelectronic applications. The key aspects of this study include quantum confinement effects, which focuses on the size-dependent changes in the electronic structure and bandgap of $CdS_{1-x}Te_x$ QDs. Quantum confinement leads to larger bandgaps and blue shifts in optical absorption and emission spectra as the QDs' diameter decreases. Also, the optical properties are studied using *ab initio* methods to calculate the absorption and emission spectra of $CdS_{1-x}Te_x$ QDs.

7.1 Introduction

Solar cells are able to convert radiation including photons to electric current. This depends on the intrinsic electronic and optical properties of materials, where the band structure and energy gap are the critical parameters of these materials. For solar cells such as CdTe, $CuInGaSe_2$, and α-Si, CdTe is regarded as the strongest candidate for large-scale manufacturing [1] and high throughput of polycrystalline solar cells. Due to their direct energy gap, 1.5 eV and high absorption coefficient, >1 × 10^4 cm^{-1}, CdTe of 1 μm thickness is quite good for absorbing around 90% of light that has energy higher than its energy gap. The Te availability is of intertest when level of production has more than 20 GW per year [2]. The electronic band structure method can be used to research phase transition of coordinate number $N_c = 4$–6-fold [3]. The goals of third-generation photovoltaics (PVs) are to increase efficiency and decrease cost using different techniques of deposition [4], which are achieving multiple energy threshold devices [5] such as hot carrier cells, up/down conversion,

doi:10.1088/978-0-7503-5704-3ch7

multiple carrier excitation, intermediate-level cells, concentrator systems, and tandem and multicolor cells.

Yum *et al* [6] have researched PbS:Hg dye-sensitized solar cells (DSCs), including quantum dot-sensitized solar cells (QDSCs), for harvesting light spectra from visible to IR. They have utilized filters for total conversion efficiency and splitting solar energy. The DSC performs 12.4% under AM1.5 G sunlight and is able to generate 9.1% of 650 nm wavelength. On the other hand, QDSC performs 5.58% for generating 3.42% of 630 nm wavelength of long-wave pass filter transmitter. QDSC and DSC-based transmitter performs 5.90% and 24.0% that guides to 13.1% efficiency via converting visible light to IR. However, Badawi *et al* [7] have investigated the performance of PVs for CdTe QDSSCs in terms of energy gap. Pre-synthesized CdTe QDs of 2.1–2.5 nm radii are prepared by direct adsorption method on TiO_2 NPs layer. The measured QDSSCs' characteristic parameters under sun illumination, AM 1.5. The overall efficiency (η) and current density (J_{sc}) values increase as CdTe QD size decreases. Where the levels of the lowest unoccupied molecular orbital shift to the level of vacuum, this increases the driving force. Moreover, the assembled response of cells to photocurrent to ONOFF cycles of illumination indicate anodic current generation. Furthermore, Emin *et al* [8] have investigated different QDSCs, which are synthesized colloidals. They have elaborated the advantages and disadvantages, working concepts, and preparation methods of different devices. The research topics integrate colloidal QDs into QDSCs, bulk heterojunction solar cells, hybrid organic–inorganic solar cells, extremely thin absorber solar cells, depleted heterojunction solar cells, and Schottky solar cells. Udipi *et al* [9] have obtained 200 nm Si QDs electron density distribution and potential energy profile. For the continuity solution equation, the efficient approximations suggested by Scharfetter and Gummel [10] have been extended to three dimensions after following up the two-dimensional approach of Selberherr *et al* [11].

The diameter dependence of materials is very interesting [12]. It is necessary to correlate nearest atom bonds and diameter dependence. By evaluating the diameter dependence of materials, it is possible to connect QD potential and dot diameter. So, to verify our QD-potential model [13], the obtained energy gaps are utilized to compute QDs' potential for diameters 60, 58, and 57 nm for $CdS_{0.25}Te_{0.75}$, $CdS_{0.5}Te_{0.5}$, and $CdS_{0.75}Te_{0.25}$ alloys, respectively, employing the full potential linearized augmented plane wave (FP-LAPW) method, in addition to investigating optical dielectric constant and refractive index optical properties utilizing specific empirical models.

7.2 Technique of $CdS_{1-x}Te_x$ alloy quantum dots

The tellurium concentration effect on optical properties of the cadmium sulfide-telluride $CdS_{1-x}Te_x$ ternary alloys ($x = 0.0, 0.25, 0.5, 0.75, 1.0$) is studied using the full potential linearized augmented plane wave method. Recently, the special quasi-random structures of Zunger's approach [14] are utilized to reproduce the alloy randomness of a few shells for a given site. The physical properties of the alloys are not affected by periodicity concept beyond a few shells that are depicted utilizing this sufficient approach.

This work has been achieved utilizing FP-LAPW method as performed in WIEN2K code [15]. The generalized gradient approximation (GGA) is attributed to exchange correlation potential [16] of total energy calculations and Engel and Vosko's generalized gradient approximation (EVGGA) formalism [17]. To exceed underestimated energy gap for using both local density approximation (LDA) and GGA [18], the EVGGA is used. This underestimation belongs to incorrect reproduction of exchange correlation energy and its charge derivative. So, the EVGGA is a modified GGA that is capable of better reproducing exchange potential at less accord with exchange energy to yield best splitting of bands. The quantities depend on an accurate exchange potential because the equilibrium bulk modulus and volume utilizing EVGGA disagree with the experimental data [18]. For the FP-LAPW method, the potential, charge density, and wave function are expanded by plane waves basis set in unit cell space (interstitial region) and spherical harmonic functions inside nonoverlapping spheres surrounding sites of atoms (muffin-tin spheres). The $l_{max} = 10$ is a wave function maximum value inside spheres of atom. The radii of the muffin-tins are 1.95 atomic units (a.u.) for both Cd and Te and 1.6 atomic units (a.u.) for S. The cut-off of plane wave for $K_{max} = 9.0/RMT$ is chosen for wave functions into interstitial region of CdS, CdTe, and $CdS_{1-x}Te_x$ alloy, while the $G_{max} = 14\ (Ryd)^{1/2}$ is a charge density of Fourier expansion. The Brillouin zone is depicted by 35 special k-points of binary mesh, and alloys of $x = 0.5$ using 64 special k-points mesh. For principal energy gaps, 220 k-points mesh is for binary and ternary alloys with $x = 0.25$ and 0.75, whereas 216 k-points mesh is for $x = 0.5$. The converged self-consistent calculations are stable within 10^{-5} Ry.

7.3 Investigations of $CdS_{1-x}Te_x$ alloy quantum dots

The four-fold coordination is a main feature of covalent semiconductor, which is attributed to low structure density and the nearest atomic neighbor is overlapped by a hybridized orbital that is known as sp^3 hybrids of tetrahedral structure. Therefore, dot diameter could tune energy gaps. The calculated direct ($\Gamma \rightarrow \Gamma$) and indirect ($\Gamma \rightarrow L$) and ($\Gamma \rightarrow X$) energy gaps using EVGGA for different dot diameters-based $CdS_{1-x}Te_x$ alloys are presented in table 7.1 in comparison with experimental [19, 20] and theoretical data [21, 22]. Our values of ($\Gamma \rightarrow \Gamma$) energy gaps are underestimated in comparison with experimental results, which is attributed to using EVGGA. So, $CdS_{1-x}Te_x$ alloys have direct energy gaps. Due to IR light detection and generation, the dot diameters-based energy gaps are important to investigate. Table 7.1 displays correlates inversely of energy gap with dot diameters as verified by figure 7.1.

The energy gaps of conduction band minimum at point X and valence band maximum at point Γ are calculated using FP-LAPW. Utilizing our model [13], the QD potential has been calculated via:

$$P_{QD} = \frac{b}{a}.\ E_{g\Gamma X}.\ 10^{-3}.\ \lambda \tag{7.1}$$

Table 7.1. Calculated principal energy gaps for QD's diameter-based $CdS_{0.75}Te_{0.25}$, $CdS_{0.5}Te_{0.5}$, and $CdS_{0.25}Te_{0.75}$ alloys (in eV) compared to other experimental and theoretical data.

QD's diameter	E_g (Γ−Γ)	E_g (Γ−X)	E_g (Γ−L)
CdS	2.359, 2.42[a] 2.361[b]	4.626	3.432
CdTe	1.368 1.8[c] 1.8[d]	3.241	2.288
$CdS_{0.75}Te_{0.25}$	0.943	0.943	2.484
61	1.067	0.949	2.609
60	1.187	0.955	2.731
59	1.306	0.960	2.850
58.5	1.422	0.964	2.967
58	1.535	0.968	3.082
57.5	0.730	2.737	2.583
$CdS_{0.5}Te_{0.5}$	0.853	2.743	2.707
63	0.974	2.748	2.829
61.5	1.092	2.753	2.949
60.5	1.208	2.758	3.066
60	1.321	2.761	3.180
59	0.650	0.650	1.538
58	0.774	0.657	1.663
$CdS_{0.25}Te_{0.75}$	0.895	0.662	1.785
65	1.013	0.667	1.940
63.5	1.129	0.670	2.021
62.5	1.242	0.675	2.136
61.5			
61			
60			

[a] Reference [19] experiment.
[b] Reference [21] theory.
[c] Reference [20] experiment.
[d] Reference [22] theory.

where constant refers to $\frac{b}{a}$ in eV^{-1} (see table 4 in reference [13]), energy gap for Γ−X in eV refers to $E_{g\Gamma X}$ and parameter refers to λ for II–VI ($\lambda = 2$), III–V ($\lambda = 4$) and group-IV ($\lambda = 6$) semiconductors in V.

There is a known correlation between pressure effect changes and QD diameter. If the QD diameter changes, the sp^3 covalent bonding is affected. Therefore, this diameter dependence of contradiction is a result of the QD's potential difference. Table 7.2 presents dot diameter-based calculated QD potential.

The separation of increasing and decreasing of QD potential is called the critical dot diameter. The correlation between transition pressure (P_t) and dot diameter is attractive to be calculated from molar free-energy difference. The Gibbs free-energy difference (ΔG_t) has dot diameter-based tetrahedral coordination and is known as

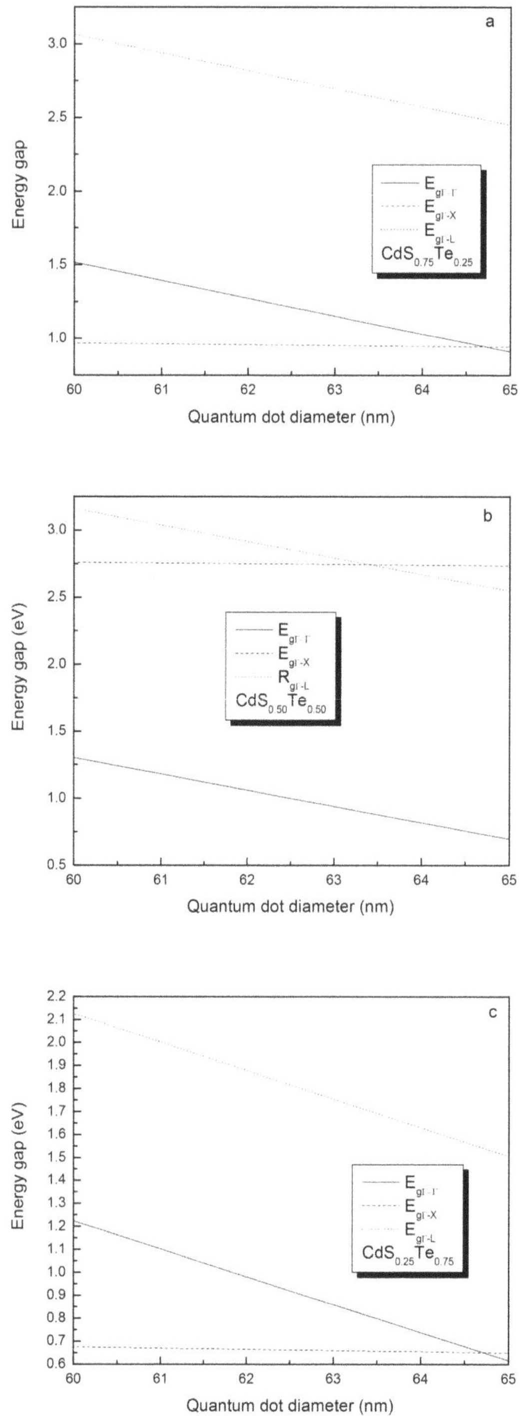

Figure 7.1. Calculated QD's diameter-based energy gaps direct ($\Gamma \rightarrow \Gamma$), and indirect ($\Gamma \rightarrow X$) and ($\Gamma \rightarrow L$) of (a) $CdS_{0.75}Te_{0.25}$, (b) $CdS_{0.5}Te_{0.5}$ and (c) $CdS_{0.25}Te_{0.75}$ alloys.

Table 7.2. Calculated dot diameter-based QD potential of $CdS_{0.75}Te_{0.25}$, $CdS_{0.5}Te_{0.5}$, and $CdS_{0.25}Te_{0.75}$ alloys in mV compared to other data in nm.

QD's diameter	P_{QD} cal.	P_{QD} [9]
$CdS_{0.75}Te_{0.25}$	0.1793	1
61	0.1804	1
60	0.1815	$\leqslant 1$
59	0.1825	
58.5	0.1833	
58	0.1840	
57.5	0.5200	
$CdS_{0.5}Te_{0.5}$	0.5212	
63	0.5222	
61.5	0.5232	
60.5	0.5240	
60	0.5247	
59	0.1236	
58	0.1248	
$CdS_{0.25}Te_{0.75}$	0.1259	
65	0.1268	
63.5	0.1276	
62.5	0.1284	
61.5		
61		
60		

$\Delta G = \Delta H - T\Delta S$ (in $kJ.mol^{-1}$), where enthalpy refers to H, temperature refers to T, and entropy refers to S. In addition, there is an inverse correlation between energies and bond lengths. Changes in the dot diameter-based QDs potential are verified by changes in the energy gap at points, $\Gamma-\Gamma$, $\Gamma-X$ and $\Gamma-L$, as presented in table 7.1. The inverse correlation between QD potential and dot diameter is displayed in table 7.2 and verified by figure 7.2. It is seen that there is a linear relationship for $CdS_{1-x}Te_x$ alloys. Therefore, QD potential fluctuations appear. Our QDs potential data are in agreement with other work [9]. This indicates that QD potential variation is a reference of electrons tunnel QDs.

The physical parameter of microscopic atomic interactions refers to refractive index n. There are two different concepts: first, the refractive index is related to local polarizability and density; and second, refractive index is related to band structure within dielectric constant [23]. As a result, real efforts have been made to relate n and E_g via specific relationships [24]. The relations of n are independent of incident photon energy and temperature. So, the different relationships discussed earlier will be elaborated. Ravindra *et al* [25] investigated a linear dependence of n:

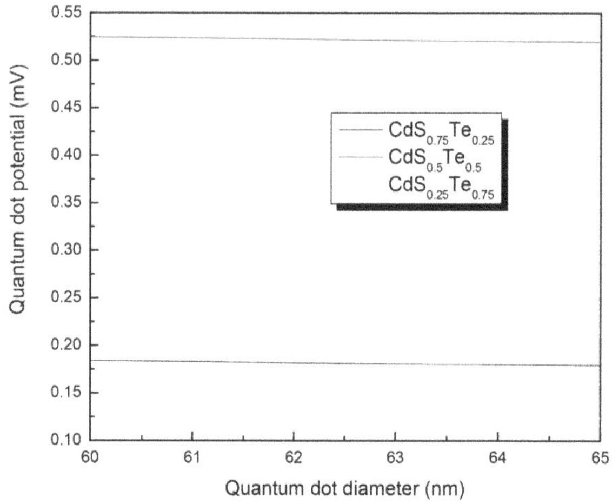

Figure 7.2. QD diameter-based QD potential for $CdS_{0.75}Te_{0.25}$, $CdS_{0.5}Te_{0.5}$, and $CdS_{0.25}Te_{0.75}$ alloys.

$$n = \alpha + \beta E_g \tag{7.2}$$

where $\alpha = 4.048$ and $\beta = -0.62\,\text{eV}^{-1}$. Herve and Vandamme [26] have suggested a relation:

$$n = \sqrt{1 + \left(\frac{A}{E_g + B}\right)^2} \tag{7.3}$$

where $A = 13.6\,\text{eV}$ and $B = 3.4\,\text{eV}$. For group II–IV, Ghosh *et al* [27] have published a relationship basing on band structure and quantum dielectric considerations of Penn [28] and Van Vechten [29]:

$$n^2 - 1 = \frac{A}{(E_g + B)^2} \tag{7.4}$$

where the average energy gaps refer to$A = 8.2E_g + 134$, $B = 0.225E_g + 2.25$, and $(E_g + B)$. Consequently, utilizing three models of n has been computed and the results are shown in figure 7.3. The calculated optical dielectric constants and refractive indices are presented in table 7.3.

The verification of optical dielectric constant ε_∞ and refractive index is done. Note that $\varepsilon_\infty = n^2$ [30]. It has been shown that Herve and Vandamme's [26] model is appropriate for calculating n because reflectivity is a critical factor to enhance solar cell photo conversion. The linear dependence of dot diameter for $CdS_{1-x}Te_x$ alloys is noticed and n for small dot diameter shifts blue–green color, which means that solar cell efficiency is affected by low reflection spectrum and high absorption. This is an impetus to investigate additional materials theoretically and experimentally to improve solar applications.

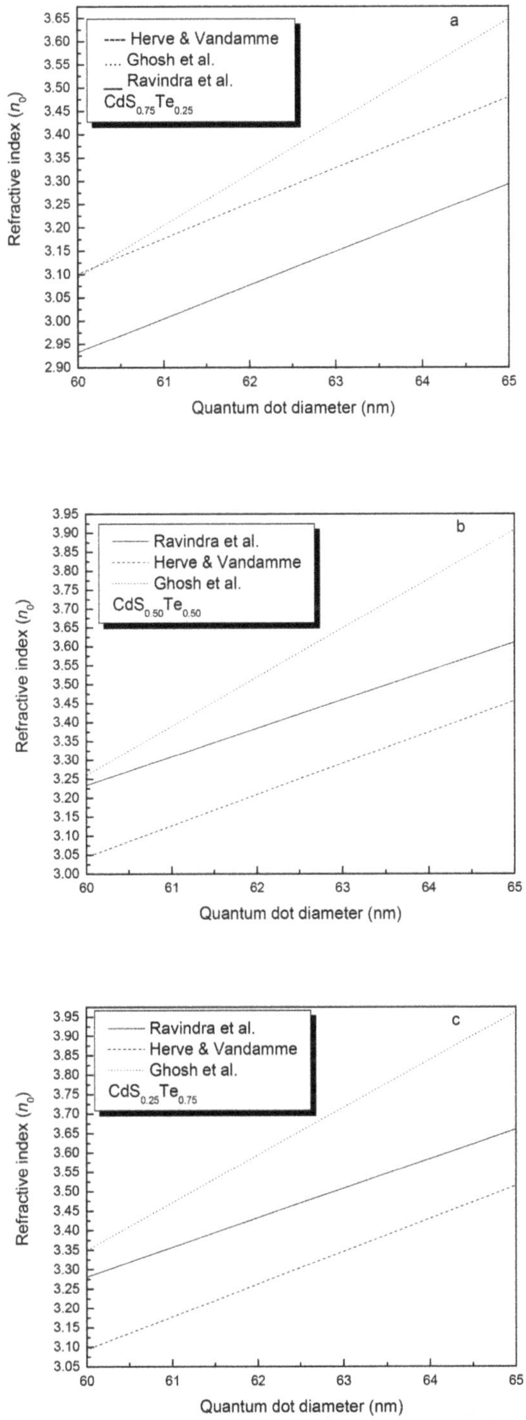

Figure 7.3. QD diameter-based refractive index (*n*) of (a) $CdS_{0.75}Te_{0.25}$ (b) $CdS_{0.5}Te_{0.5}$ and (c) $CdS_{0.25}Te_{0.75}$ alloys.

Table 7.3. Calculated dot diameter-based refractive indices of $CdS_{0.75}Te_{0.25}$, $CdS_{0.5}Te_{0.5}$, and $CdS_{0.25}Te_{0.75}$ alloys utilizing Ravindra *et al's* [25], Herve and Vandamme's [26], and Ghosh *et al's* [27] models corresponding to optical dielectric constant.

QD's diameter	n	ε_∞
CdS	2.589[a] 2.567[b] 2.611[c] 2.38[d]	6.702[a] 2.589[b] 2.817[c]
CdTe	3.345[a] 3.1627[b] 3.439[c] 2.7[d]	11.189[a] 9.998[b] 11.826[c]
$CdS_{0.75}Te_{0.25}$	3.46[a] 3.28[b] 3.63[c]	11.97[a] 10.75[b] 13.17[c]
61	3.38[a] 3.20[b] 3.50[c]	11.42[a] 10.24[b] 12.25[c]
60	3.31[a] 3.12[b] 3.38[c]	10.95[a] 9.73[b] 11.42[c]
59	3.23[a] 3.05[b] 3.28[c]	10.43[a] 9.30[b] 10.75[c]
58.5	3.16[a] 2.99[b] 3.18[c]	9.98[a] 8.94[b] 10.11[c]
58	3.09[a] 2.93[b] 3.09[c]	9.54[a] 8.58[b] 9.54[c]
57.5	3.59[a] 3.44[b] 3.89[c]	12.88[a] 11.83[b] 15.13[c]
$CdS_{0.5}Te_{0.5}$	3.51[a] 3.35[b] 3.73[c]	12.32[a] 11.22[b] 13.91[c]
63	3.44[a] 3.26[b] 3.60[c]	11.83[a] 10.62[b] 12.96[c]
61.5	3.37[a] 3.18[b] 3.47[c]	11.35[a] 10.11[b] 12.04[c]
60.5	3.29[a] 3.11[b] 3.36[c]	10.82[a] 9.67[b] 11.28[c]
60	3.22[a] 3.04[b] 3.26[c]	10.36[a] 9.24[b] 10.62[c]
59	3.64[a] 3.50[b] 3.91[c]	13.24[a] 12.25[b] 15.28[c]
58	3.56[a] 3.40[b] 3.83[c]	12.67[a] 11.56[b] 14.66[c]
$CdS_{0.25}Te_{0.75}$	3.49[a] 3.32[b] 3.69[c]	12.18[a] 11.02[b] 13.61[c]
65	3.41[a] 3.23[b] 3.55[c]	11.62[a] 10.43[b] 12.60[c]
63.5	3.34[a] 3.16[b] 3.44[c]	11.15[a] 9.98[b] 11.83[c]
62.5	3.27[a] 3.09[b] 3.33[c]	10.69[a] 9.54[b] 11.08[c]
61.5		
61		
60		

[a] Reference [25].
[b] Reference [26].
[c] Reference [27].
[d] Reference [25] experiment.

FP-LAPW has provided an excellent method for calculating and investigating optoelectronic properties of low reflectivity of $CdS_{1-x}Te_x$ alloys proving that 60, 58, and 57 nm for $CdS_{0.25}Te_{0.75}$, $CdS_{0.5}Te_{0.5}$, and $CdS_{0.75}Te_{0.25}$ alloys, respectively, while Herve and Vandamme's model is appropriate for photovoltaics to develop novel trends of alloys and new horizons for QDs.

Problems

1. What is the ab initio method? And, how is it used to investigate the optical properties of materials such as $CdS_{1-x}Te_x$ alloys?
2. How does the diameter of QDs in $CdS_{1-x}Te_x$ alloys affect their optical properties? And, how can this be investigated using the ab initio method?

3. What are some of the challenges associated with using the ab initio method to investigate the optical properties of $CdS_{1-x}Te_x$ alloys under QDs diameter effect? And, how can they be addressed?
4. What are some of the potential applications of $CdS_{1-x}Te_x$ alloys with tunable optical properties? And, how can the ab initio method be used to optimize their performance?
5. How does the composition of $CdS_{1-x}Te_x$ alloys affect their optical properties? And, how can the ab initio method be used to model and understand these effects? Note: $CdS_{1-x}Te_x$ alloys refer to materials that are composed of a mixture of cadmium sulfide (CdS) and cadmium telluride (CdTe) with varying ratios of each component, denoted by the parameter x.

References

[1] Hsu H-R, Hsu S-C and Liu Y S 2012 *Sol. Energy* **86** 48

[2] Boucher J W, Miller D W, Warren C W, Cohen J D, McCandless B E, Heath J T, Lonergan M C and Boettcher S W 2014 *Sol. Energy Mater. Sol. Cells* **129** 57

[3] Zhang S B and Cohen M L 1987 *Phys. Rev.* B **35** 7604

[4] Green M A 2003 *Third Generation Photovoltaics: Ultra-High Efficiency at Low Cost* (Berlin: Springer)

[5] Nelson J 2003 *The Physics of Solar Cells* (London: Imperial College Press)

[6] Yum J-H, Lee J-W, Kim Y, Humphry-Baker R, Park N-G and Grätzel M 2014 *Sol. Energy* **109** 183

[7] Badawi A, Al-Hosiny N, Abdallah S, Negm S and Talaat H 2013 *Sol. Energy* **88** 137

[8] Emin S, Singh S P, Han L, Satoh N and Islam A 2011 *Sol. Energy* **85** 1264

[9] Udipi S, Vasileska D and Ferry D K 1996 *Superlattices Microstruct.* **20** 343

[10] Scharfetter D L and Gummel H K 1969 *IEEE Trans. Electron Devices* **ED-16** 64

[11] Selberherr S, Shutz A and Potzl H W 1980 *IEEE Trans. Electron Devices* **ED-27** 1540

[12] Al-Douri Y, Baaziz H, Charifi Z, Khenata R, Hashim U and Al-Jassim M 2012 *Renew. Energy* **45** 232

[13] Al-Douri Y 2009 Quantum dot modeling of semiconductors in: A H Reshak ed *Advances in Condensed Matter Physics* (Kerala: Research Signpost)

[14] Zunger A, Wei S -H, Ferreira L G and Bernard E 1990 *Phys. Rev. Lett.* **65** 353

[15] Blaha P, Schwarz K, Madsen G K H, Kvasnicka D and Luitz J 2001 *WIEN2K* (Wien, Austria: Technische Universitat)

[16] Perdew J P, Burke S and Ernzerhof M 1996 *Phys. Rev. Lett.* **77** 3865

[17] Engel E and Vosko S H 1993 *Phys. Rev.* B **47** 13164

[18] Dufek P, Blaha P and Schwarz K 1994 *Phys. Rev.* B **50** 7279

[19] Lincot D and Hodes G 2006 *Chemical Solution Deposition of Semiconducting and Non-Metallic Films* (Pennington, NJ: Electrochemical Society)

[20] Tsidilkovski I M 1982 *Band Structure of Semiconductors* (Oxford: Pergamon Press)

[21] Feng Y P, Teo K L, Li M F, Poon H C, Ong C K and Xia J B 1993 *J. Appl. Phys.* **74** 3948

[22] Al-Douri Y, Abid H, Zaoui A and Aourag H 2001 *Physica* B **301** 295

[23] Balzaretti N M and da Jornada J A H 1996 *Solid State Commun.* **99** 943

[24] Al-Douri Y, Feng Y P and Huan A C H 2008 *Solid State Commun.* **148** 521

[25] Ravindra N M, Auluck S and Srivastava V K 1979 *Phys. Status Solidi* B **93** K155

[26] Herve P J L and Vandamme L K J 1995 *J. Appl. Phys.* **77** 5476
[27] Ghosh D K, Samanta L K and Bhar G C 1984 *Infrared Phys.* **24** 34
[28] Penn D R 1962 *Phys. Rev.* **128** 2093
[29] Van Vechten J A 1969 *Phys. Rev.* **182** 891
[30] Samara G A 1983 *Phys. Rev.* B **27** 3494

Chapter 8

First-principles calculations for optical investigations of PbX (X = S, Te) compounds under quantum dots diameter effect

This chapter explores the optical properties of lead chalcogenide quantum dots (QDs; PbS and PbTe) influenced by their size using first-principles methods. This research aims to understand how quantum confinement affects the electronic and optical characteristics of these semiconductor nanocrystals, which are significant for applications in optoelectronics and photonics. The key elements of the study include quantum confinement, which examines how the diameter of PbS and PbTe QDs influences their electronic structure and optical properties. As the size of the QDs decreases, quantum confinement leads to an increase in the bandgap and shifts in the optical absorption and emission spectra. Meanwhile, optical properties determine the absorption and emission spectra of PbS and PbTe QDs. These calculations reveal size-dependent changes in the optical transitions, which are critical for designing materials with specific optical functionalities.

8.1 Introduction

Solid-state physics regards IV–VI materials as interesting, including PbS and PbTe compounds that have narrow energy gap for about 0.5 eV, positive temperature coefficient $dE_g = dT$, and high mobility [1], leading to successful application in optoelectronics [2]. So, it is beneficial to use total energy calculations to investigate coordination of N_c = 4–6-fold phase transition [3]. Research aims to discover alternatives that are cost effective and have increased efficiency [4]. Different approaches are available to fulfill the required technologies, such as multiple energy threshold devices [5], hot carrier cells, up/down conversion, multiple carrier excitation, intermediate-level cells, intermediate-level cells, concentrator systems, and multicolor or tandem cells [6].

Billaud and Truong [7] have calculated the Lamb shift of ground state of QDs in approximation of effective mass. This seems a significant step to detect small

doi:10.1088/978-0-7503-5704-3ch8

synthesized QDs. The Casimir effect has been suggested by [8] for observing. Meanwhile, Thu and Voskoboynikov [8] have computed lowest electronic energy confined in InAs/GaAs double lens-shaped QDs molecule under magnetic-field effect. Based on effective one electronic-band Hamiltonian approximation of three dimensions, electronic energies are calculated by Comsol MultiPhysics package for the nonlinear iterative method. This allows us to study QDs molecule under magnetic-field effect. This study has shown that diamagnetic shifts of electronic levels of energy are anisotropic and nonuniform. So, the authors have presented electronic states of dynamically manipulate by varying and changing magnitude and direction of magnetic field. Furthermore, Lam and Ng [9] have researched biotags of different QD sizes for emitting different colors that are used to study artificial atom applications, such as quantum transport energies, electron spins, and discrete energies. The self-assembled QDs are grown on wetting layers of a few monolayers capped with strain-reduction layers covering QDs for stabilization. The authors have investigated InAS/GaAs self-assembled QDs modeled with wetting layers between QDs and substrate aside, and strain-reducing capping layers above QDs on another side, where they have presented new interfacial layer model between QDs and capping layer to research effective electronic and mechanical properties utilizing deformation potential theory and finite element method. Udipi *et al* [10] have investigated 200 nm Si QDs of electron density distribution and potential energy profile. For continuity of the equation of the solution, the difference approximation of efficiency suggested by Scharfetter and Gummel [11] has been extended to three dimensions, following up the two-dimensional approach of Selberherr *et al* [12].

Further investigation is necessary to obtain more information about the diameter dependence of materials [13]. It looks very important to relate nearest atoms bonds and diameter dependence behavior. Controlling diameter dependence evaluation is attractive to connect QDs effect to QDs potential. Therefore, this trend has been followed up to validate our QD potential model [14], so the calculated energy gaps are utilized to calculate QD potential and predicting new QD materials. The calculated dot diameter-based QDs potential [14] of 65 and 60 nm for PbTe and PbS materials, respectively, are verified successfully by employing the full potential linearized augmented plane wave (FP-LAPW) method, in addition to an investigation of the optical dielectric constant and refractive index.

8.2 Technique of PbX (X = S, Te) alloy quantum dots

The FP-LAPW implemented into WIEN2K code [15] has been employed to calculate the different properties of materials. The generalized gradient approximation (GGA) [16] within exchange correlation potential is consumed for total energy calculations, in addition to Engel and Vosko's generalized gradient approximation (EVGGA) formalism [17] and modified Becke Johnson (mBJ) [18] for calculation. The EVGGA and mBJ are utilized to overcome the underestimation of LDA and GGA [19] because neither approximation correctly reproduces exchange correlation energy and its charge derivative. So, the EVGGA is improved in mBJ, which is better at reproducing the exchange potential at less agreement in exchange energy to yield better splitting of bands [20]. For FP-LAPW, the potential, charge density, and wave function are expanded by spherical harmonic functions into nonoverlapping spheres surrounding sites of atoms (muffin-tin spheres) and by basis sets of plane waves in unit cell space (interstitial region). The maximal $l_{max} = 8$ is a wave function expansion into spheres of atoms. The radii of muffin-tin are 2.0 atomic units (a.u.) for Te, S and Pb. The $K_{max} = 8.0/RMT$ if plane wave cut-off is chosen for wave functions expansion in interstitial region of PbTe and PbS, and $G_{max} = 14\ (Ryd)^{1/2}$ is a Fourier expansion of charge density. The irreducible Brillouin zone is depicted by 10 special k-points mesh of binary semiconductors. The converged self-consistent total energy indicated for a stable system is 10^{-5} Ry. The FP-LAPW has proven to be an accurate method for studying the electronic properties of materials [21–24].

8.3 Investigation of PbX (X = S, Te) QDs

The four-fold coordination is a main feature of covalent semiconductors. The low density of structure and overlapped nearest neighbors of atoms bound by hybridized orbitals, which are known as sp^3 hybrids of tetrahedral structure. So, it is possible to tune the energy gaps using the QDs' diameters. The calculated direct ($\Gamma \rightarrow \Gamma$) and indirect ($\Gamma \rightarrow X$) and ($\Gamma \rightarrow L$) energy gaps using EVGGA and mBJ for PbTe and PbS at different QDs' diameters are displayed in table 8.1 compared with

Table 8.1. Calculated QD's diameter-based energy gaps of PbS and PbTe in eV compared to experimental and theoretical data.

QD's diameter	E_g (Γ−Γ)	E_g (Γ−X)	E_g (Γ−L)
PbS	1.2448[C1] 1.1888[C2] 0.286[a] 0.989[b] 0.069[c]	7.1691[C1] 6.786[C2]	5.2538[C1] 4.6518[C2]
Bulk	C_1 C_2	C_1 C_2	C_1 C_2
60	7.17 7.74	7.16 6.83	5.25 5.61
61	6.75 7.19	7.12 6.81	5.11 5.38
62	6.42 6.86	6.82 6.79	4.98 5.28
63	6.19 6.58	6.57 6.78	4.87 5.18
64	5.90 6.34	6.26 6.67	4.75 5.08
65	5.69 6.14	6.04 6.45	4.65 4.99
PbTe	1.2948[C1] 1.185[C2] 0.19[a] 0.5704[b] 0.032[c]	5.6827[C1] 5.1268[C2]	4.0139[C1] 3.8054[C2]
Bulk	C_1 C_2	C_1 C_2	C_1 C_2
65	5.34 5.52	5.68 5.12	4.01 4.20
67	4.99 5.10	5.61 5.25	3.91 3.98
68	4.73 4.85	5.52 5.28	3.82 3.90
69	4.52 4.64	5.44 5.30	3.73 3.83
70	4.34 4.47	5.20 5.32	3.65 3.75
71	4.19 4.33	5.00 5.16	3.57 3.68

[a] Reference [25] experiment.
[b] Reference [26] theory.
[c] Reference [27] theory.
[C1] calculated by EVGGA.
[C2] calculated by mBJ.

experimental [25] and theoretical data [26, 27]. Our (Γ → Γ) energy gaps are overestimated in comparison with others, which is attributed to utilizing EVGGA and mBJ, so, both PbTe and PbS are classified as indirect energy gap semiconductors. Due to utilizing infrared light detection and generation, QD diameter-based energy gap variations are regard as critical for this study. Table 8.1 displays an inverse correlation of energy gaps correlates with QD diameter, as verified by figure 8.1.

The energy gaps between conduction band minimum at point X and valence band maximum at point Γ are calculated using FP-LAPW. Our model [14] has been utilized to evaluate QDs' potential according to:

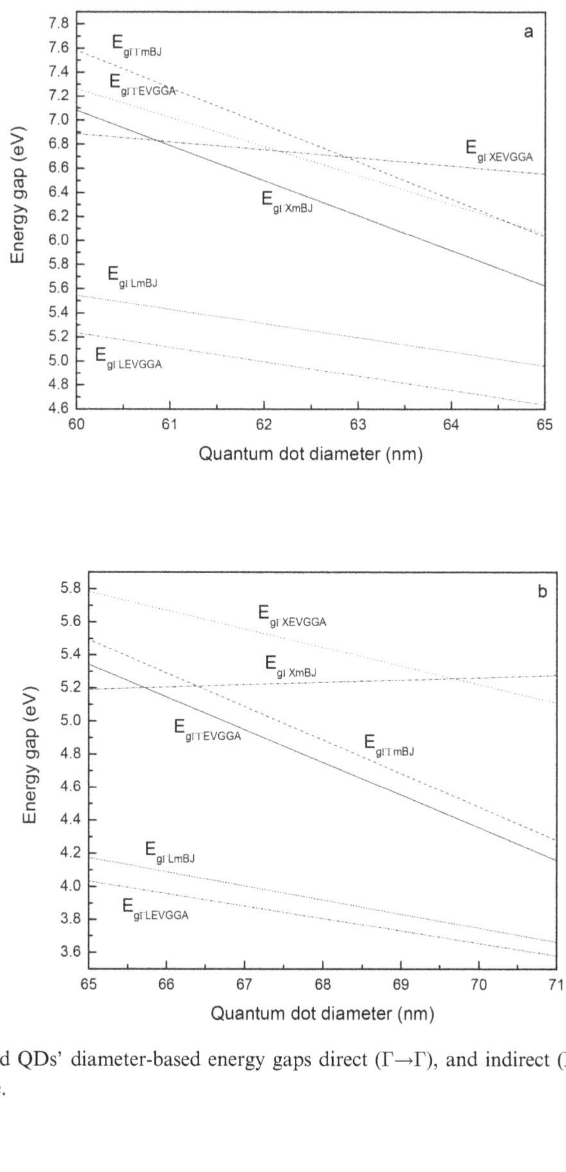

Figure 8.1. Calculated QDs' diameter-based energy gaps direct (Γ→Γ), and indirect (Γ→X) and (Γ→L) for (a) PbS and (b) PbTe.

Table 8.2. Calculated QD's diameter-based on the QD's potential of PbS and PbTe in mV compared to other work.

QD's diameter	P_{QD} cal. × 10^{-3}	P_{QD} [10]
PbS	C_1 C_2	1
60	1.36 1.299	⩽ 1
61	1.35 1.293	
62	1.29 1.290	
63	1.24 1.28	
64	1.18 1.26	
65	1.14 1.22	
PbTe	1.07 0.972	
65	1.06 0.997	
67	1.04 1.003	
68	1.03 1.007	
69	0.98 1.010	
70	0.95 0.980	
71		

C_1: calculated by EVGGA, C_2: calculated by mBJ.

$$P_{QD} = \frac{b}{a}.\ E_{g\Gamma X}.\ 10^{-3}.\ \lambda \tag{8.1}$$

where constant refers $\frac{b}{a}$ is in eV^{-1} (see table 4 in Ref. 14), energy gap refers to $E_{g\Gamma X}$ along $\Gamma - X$ in eV, and parameter refers to λ for II–VI ($\lambda = 2$), III–V ($\lambda = 4$) and group-IV ($\lambda = 6$) semiconductors in V.

There is a known correlation between pressure effect changes and the QD's diameter. If the QD's diameter changes, the sp^3 covalent bond is affected. Distinguishing the QD's diameter is a result of the QD's potential difference. Table 8.2 presents calculated the QD's diameter-based potential.

The separation between decreasing and increasing QDs potential is a critical QD diameter. The correlation between transition pressure (P_t) and diameter dependence is interesting for calculating the molar free energies difference. The Gibbs free-energy difference (ΔG_t) at QDs' diameter dependence is given by $\Delta G = \Delta H - T\Delta S$ (in kJ mol^{-1}), where enthalpy refers to H, temperature refers to T, and entropy refers to S. In addition, there is an inverse correlation between energies and bond lengths. The QDs' diameter-based potential change is verified by energy gap changes at $\Gamma - \Gamma$, $\Gamma - X$ and $\Gamma - L$, as displayed in table 8.1. The inverse correlation between the QDs' potential and QDs' diameter is presented in table 8.2 and verified by figure 8.2. It appears as linear and nonlinear relationships of PbS and PbTe compounds, respectively. As a result, QDs' potential fluctuations appear. Our QDs' potential data agree with previous work [10]. This indicates that QDs' potential variation is a reference of electron tunnel QDs.

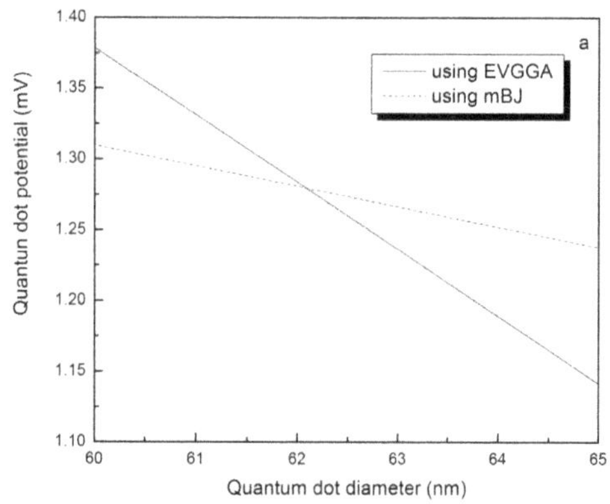

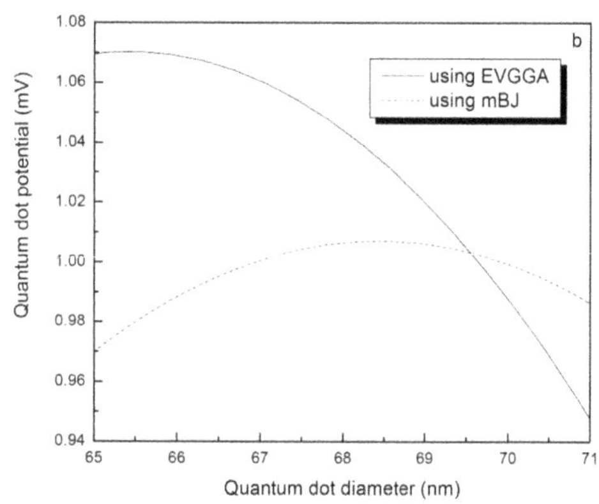

Figure 8.2. QDs diameter-based QDs potential of (a) PbS and (b) PbTe.

The optical properties of microscopic atomic interactions refer to refractive index n. Theoretically, there are two approaches: first, the n is related to local polarizability and density; and second, n is related to energy gaps through dielectric constant [28]. As a result, many relationship attempts are established to relate n and E_g [29]. These relations are independent of incident photon energy and temperature. Different relationships of n and E_g will be elaborated. Ravindra *et al* [30] have researched a linear relationship of n and E_g:

$$n = \alpha + \beta E_g \tag{8.2}$$

where $\alpha = 4.048$ and $\beta = -0.62\ \text{eV}^{-1}$. Herve and Vandamme [31] have proposed an empirical relation as follows:

$$n = \sqrt{1 + \left(\frac{A}{E_g + B}\right)^2} \tag{8.3}$$

where $A = 13.6\,\text{eV}$ and $B = 3.4\,\text{eV}$. For group II–IV semiconductors, Ghosh *et al* [32] have published an empirical relationship based on the band structure and quantum dielectric considerations of Penn [33] and Van Vechten [34]:

$$n^2 - 1 = \frac{A}{(E_g + B)^2} \tag{8.4}$$

where $A = 8.2E_g + 134$, $B = 0.225E_g + 2.25$, and $(E_g + B)$ refers to appropriate average energy gaps. So, utilizing the three models of n is necessary for optical studies. The outcomes are illustrated in figure 8.3. The calculated optical dielectric constants and n researched and presented in table 8.3.

It is confirmed that refractive index-based optical dielectric constant ε_∞ is a fact. Note that $\varepsilon_\infty = n^2$ [37]. It is confirmed that Ghosh *et al's* [32] model and EVGGA

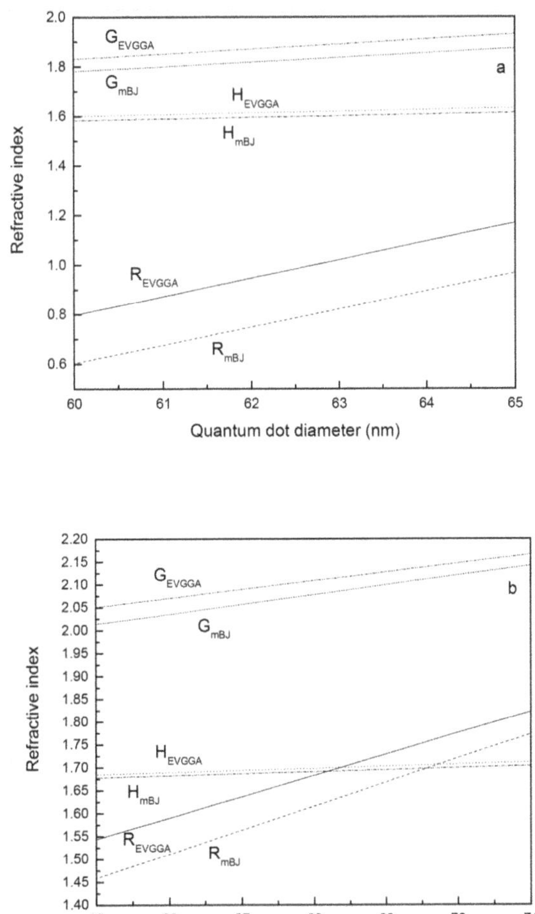

Figure 8.3. QDs diameter-based refractive index (n) of (a) PbS and (b) PbTe. R: Ravindra *et al* [30], H: Herve and Vandamme [31], G: Ghosh *et al* [32].

Table 8.3. Calculated QDs' diameter-based refractive indices for PbS and PbTe using Ravindra *et al's* [30], Herve and Vandamme's [31], and Ghosh *et al's* [32] models corresponding to optical dielectric constant.

QD's diameter	n	ε_∞
PbS	3.27[a,1] 3.31[a,2] 3.09[b,1] 3.12[b,2] 4.26[c,1]	10.69[a,1] 10.95[a,2] 9.54[b,1] 9.73[b,2] 18.14[c,1]
Bulk	4.30[c,2] 4.16[d]	18.49[c,2]
60	0.79[a,1] 0.56[a,2] 1.60[b,1] 1.58[b,2] 1.83[c,1]	0.62[a,1] 0.31[a,2] 2.56[b,1] 2.49[b,2] 3.34[c,1]
61	1.77[c,2]	3.13[c,2]
62	0.87[a,1] 0.71[a,2] 1.61[b,1] 1.59[b,2] 1.85[c,1]	0.75[a,1] 0.50[a,2] 2.59[b,1] 2.52[b,2] 3.42[c,1]
63	1.81[c,2]	3.27[c,2]
64	0.96[a,1] 0.77[a,2] 1.61[b,1] 1.60[b,2] 1.87[c,1]	0.92[a,1] 0.59[a,2] 2.59[b,1] 2.56[b,2] 3.49[c,1]
65	1.82[c,2]	3.31[c,2]
PbTe	1.02[a,1] 0.83[a,2] 1.62[b,1] 1.60[b,2] 1.89[c,1]	1.04[a,1] 0.68[a,2] 2.62[b,1] 2.56[b,2] 3.57[c,1]
Bulk	1.84[c,2]	3.38[c,2]
65	1.10[a,1] 0.89[a,2] 1.63[b,1] 1.61[b,2] 1.91[c,1]	1.21[a,1] 0.79[a,2] 2.65[b,1] 2.59[b,2] 3.64[c,1]
67	1.85[c,2]	3.42[c,2]
68	1.16[a,1] 0.95[a,2] 1.63[b,1] 1.61[b,2] 1.93[c,1]	1.34[a,1] 0.90[a,2] 2.65[b,1] 2.65[b,2] 3.72[c,1]
69	1.87[c,2]	3.49[c,2]
70	3.24[a,1] 3.31[a,2] 3.06[b,1] 3.13[b,2] 6.22[c,1]	10.49[a,1] 10.95[a,2] 9.36[b,1] 9.79[b,2] 38.68[c,1]
71	6.31[c,2] 5.98[e]	39.81[c,2]
	1.56[a,1] 1.44[a,2] 1.69[b,1] 1.68[b,2] 2.06[c,1]	2.43[a,1] 2.07[a,2] 2.85[b,1] 2.82[b,2] 4.24[c,1]
	2.01[c,2]	4.04[c,2]
	1.62[a,1] 1.58[a,2] 1.69[b,1] 1.685[b,2] 2.08[c,1]	2.62[a,1] 2.49[a,2] 2.85[b,1] 2.82[b,2] 4.32[c,1]
	2.06[c,2]	4.24[c,2]
	1.67[a,1] 1.63[a,2] 1.695[b,1] 1.69[b,2] 2.10[c,1]	2.78[a,1] 2.65[a,2] 2.87[b,1] 2.85[b,2] 4.41[c,1]
	2.08[c,2]	4.41[c,2]
	1.73[a,1] 1.67[a,2] 1.70[b,1] 1.695[b,2] 2.13[c,1]	2.99[a,1] 2.78[a,2] 2.89[b,1] 2.85[b,2] 4.53[c,1]
	2.10[c,2]	4.41[c,2]
	1.78[a,1] 1.72[a,2] 1.71[b,1] 1.70[b,2] 2.15[c,1]	3.16[a,1] 2.95[a,2] 2.92[b,1] 2.89[b,2] 4.62[c,1]
	2.12[c,2]	4.49[c,2]
	1.83[a,1] 1.76[a,2] 1.715[b,1] 1.705[b,2] 2.17[c,1]	3.34[a,1] 3.09[a,2] 2.94[b,1] 8.70[b,2] 4.70[c,1]
	2.14[c,2]	4.57[c,2]

[a] Reference. [30].
[b] Reference [31].
[c] Reference [32].
[d] Reference [35] experiment.
[e] Reference [36] experiment.
[1] Calculated by EVGGA.
[2] Calculated by mBJ.

are suitable for improving photon conversion and photovoltaic applications. The QDs' diameter-based refractive index shows a linear dependence of PbS and PbTe, and shifting towards blue–green color, which means that low reflection and high absorption are attributed to increased efficiency solar cells.

The FP-LAPW is a good method for calculating electronic properties and investigating optical properties of low reflectivity y of PbS and PbTe to present evidence that 65 and 60 nm of PbTe and PbS, respectively, via Ghosh *et al's* model that are suitable for applications in solar cells and it is expected that further materials will be realized by QDs. Finally, EVGGA is appropriate for QD potential.

Problems

1. What are first-principles calculations? And, how can they be used to investigate the optical properties of materials such as PbX (X = S, Te) compounds under the QD diameter effect?
2. How does the diameter of QDs in PbX (X = S, Te) compounds affect their optical properties? And, how can this be investigated using first-principles calculations?
3. What are some of the challenges associated with using first-principles calculations to investigate the optical properties of PbX (X = S, Te) compounds under QDs diameter effect? And, how can they be addressed?
4. What are some of the potential applications of PbX (X = S, Te) compounds with tunable optical properties? And, how can first-principles calculations be used to optimize their performance?
5. How does the composition of PbX (X = S, Te) compounds affect their optical properties? And, how can first-principles calculations be used to model and understand these effects? Note: PbX (X = S, Te) compounds refer to materials that are composed of lead sulfide (PbS) and lead telluride (PbTe) with varying ratios of each component.

References

[1] Agrawal G P and Dutta N K 1993 *Semiconductor Lasers* (New York: Van Nostrand Reinhold)

[2] Nair P K, Ocampo M, Fernandez A and Nair M T S 1990 *Sol. Energy Mater.* **20** 235

[3] Zhang S B and Cohen M L 1987 *Phys. Rev.* B **35** 7604

[4] Green M A 2003 *Third Generation Photovoltaics: Ultra-High Efficiency at Low Cost* (Berlin: Springer)

[5] Nelson J 2003 *The Physics of Solar Cells* (London: Imperial College Press)

[6] Conibeer G 2007 *Mater. Today* **10** 42

[7] Billaud B and Truong T-T 2011 *Comput. Mater. Sci.* **50** 998

[8] Thu L M and Voskoboynikov O 2010 *Comput. Mater. Sci.* **49** S281

[9] Lam A W and Ng T Y 2010 *Comput. Mater. Sci.* **49** S54

[10] Udipi S, Vasileska D and Ferry D K 1996 *Superlattices Microstruct.* **20** 343

[11] Scharfetter D L and Gummel H K 1969 *IEEE Trans. Electron Devices* **ED-16** 64

[12] Selberherr S, Shutz A and Potzl H W 1980 *IEEE Trans. Electron Devices* **ED-27** 1540

[13] Al-Douri Y, Baaziz H, Charifi Z, Khenata R, Hashim U and Al-Jassim M 2012 *Renew. Energy* **45** 232

[14] Al-Douri Y 2009 Quantum dot modeling of semiconductors in: A H Reshak ed *Advances in Condensed Matter Physics* (Kerala, India: Research Signpost)

[15] Blaha P, Schwarz K, Madsen G K H, Kvasnicka D and Luitz J 2001 *WIEN2K* (Wien, Austria: Technische Universitat)
[16] Perdew J P, Burke S and Ernzerhof M 1996 *Phys. Rev. Lett.* **77** 3865
[17] Engel E and Vosko S H 1993 *Phys. Rev.* B **47** 13164
[18] Becke A D and Johnson E R 2006 *J. Chem. Phys.* **124** 221101
[19] Dufek P, Blaha P and Schwarz K 1994 *Phys. Rev.* B **50** 7279
[20] Reshak A H 2015 *RSC Adv.* **5** 22044
[21] Melosh N, Boukai A, Diana F, Gerardot B, Badolato A, Petroff P and Heath J R 2003 *Science* **300** 112
[22] Goicoechea J, Zamarreñoa C R, Matiasa I R and Arregui F J 2007 *Sensors Actuators* B **126** 41
[23] Waldner J-B 2007 *Nanocomputers and Swarm Intelligence* (London: Wiley)
[24] Cao M, Wang L, Zhang Q, Zhang H, Zhong S and Chen J 2023 *Mater. Today Commun.* **36** 106677
[25] Strehlow W H and Cook E L 1973 *J. Phys. Chem. Ref. Data* **2** 163
[26] Rached D, Rabah M, Benkhettou N, Driz M and Soudini B 2003 *Physica* B **337** 394
[27] Delin A, Ravindran P, Eriksson O and Wills J M 1998 *Int. J. Quantum Chem.* **69** 349
[28] Balzaretti N M and da Jornada J A H 1996 *Solid State Commun.* **99** 943
[29] Al-Douri Y, Feng Y P and Huan A C H 2008 *Solid State Commun.* **148** 521
[30] Ravindra N M, Auluck S and Srivastava V K 1979 *Phys. Status Solidi* B **93** K155
[31] Herve P J L and Vandamme L K J 1995 *J. Appl. Phys.* **77** 5476
[32] Ghosh D K, Samanta L K and Bhar G C 1984 *Infrared Phys.* **24** 34
[33] Penn D R 1962 *Phys. Rev.* **128** 2093
[34] Van Vechten J A 1969 *Phys. Rev.* **182** 891
[35] Zemel J N, Jensen J D and Schoolar R B 1965 *Phys. Rev.* **140** A330
[36] Weiting F and Yixun Y 1990 *Infrared Phys.* **30** 371
[37] Samara G A 1983 *Phys. Rev.* B **27** 3494

Chapter 9

Chalcogenides-based quantum dots: optical investigation using first-principles calculations

This chapter focuses on the optical properties of quantum dots (QDs) composed of chalcogenide materials (e.g., CdS, CdSe, CdTe, PbS, and PbTe) using first-principle computational methods. This research aims to understand how quantum confinement effects influence the electronic and optical behavior of these semiconductor nanocrystals. The key aspects of the study include quantum confinement effects, which examine how reducing the size of chalcogenide QDs alters their bandgap and energy levels. The quantum confinement effect leads to discrete energy states and size-dependent shifts in the optical absorption and emission spectra. Meanwhile, the optical properties are used to analyze the absorption and emission spectra of various chalcogenide QDs. The results highlight how the optical transitions change with QD size and composition, which is crucial for tailoring materials for specific optical applications.

9.1 Introduction

The solid-state physics of IV–VI semiconductors has acquired a lot of interest. PbTe, PbSe, and PbS are IV–VI semiconductors, which have high mobility, positive temperature coefficient $\alpha_T = \frac{\partial E_g}{\partial T}$, and narrow energy gaps <0.5 eV [1]. Their obvious application is in optoelectronics [2]. The total energy calculation-based computational technique is useful for researching phase transition of coordination $N_c = 4$–6-fold [3]. It is well known that the new generation of photovoltaics (PVs) aims to increase efficiency and be cost-effective, while keeping the environmental and economic factors of technological techniques [4]. In addition, the hot carrier cells, up/down conversion, multiple carrier excitation, intermediate-level cells, concentrator systems, and multicolor and tandem cells are attractive approaches for energy threshold devices [5].

doi:10.1088/978-0-7503-5704-3ch9

Luo *et al* [6] have prepared water-soluble Ge:CdS QDs via one-pot technique using probe of fluorescent for live MCF-7 cell labeling. The QDs' optical properties and synthetic parameters are researched, including a discussion of the mechanism parameters. Moreover, the CdS QDs toxicity decreases as Ge-doped increases, improving CdS QDs biocompatibility. Then, folic acid-doped QDs are used as fluorescent probes to widen the biomedical applications of QDs. Meanwhile, El-Rabaie [7] have studied PbTe QDs growth embedding into a new matrix of fluorogermante glass. The absorption illustrates a big shift of blue color, implying synthesized PbTe QDs and reflects the goof effect of quantum confinement. So, the size of nanoparticle is to 7.3, 6.1, 3.6, and 3.2 at 500 °C for 180, 120, 60, and 30 min, respectively. The energy gap decreases from 2.52–1.95 eV as the size of the nanoparticles increases. In addition, x-ray diffraction (XRD) schemes have confirmed a cubic structure of PbTe nanocrystals. Furthermore, Strong *et al* [8] have investigated GaAs diodes with and without InAS; the QDs are irradiated to investigate induced defects. GaAs QDs pn-junction diodes are analyzed by measuring deep level transient spectroscopy and capacitance–voltage. The accumulation of carriers is noticed in GaAS QDs at a designed depth of QDs via measuring capacitance–voltage. In QDs after 1 MeV irradiation, QD–E3 (E_C–0.28 eV), QD–E4 (E_C–0.49 eV) and QD–EL2 (E_C–0.72 eV) defects are similar to baseline, which are noticed even though the trap density is dissimilar to baseline. The QDs show QD–E4 defect higher density and QD–E3 lower density, but the QD–EL2 defect seems to be unaffected by irradiation. Udipi *et al* [9] have submitted 200 nm Si QDs electron density distribution and potential energy profile results. For the continuity equation of the solution, the difference approximation efficiency suggested by Scharfetter and Gummel [10] has been extended to three dimensions following up on the two-dimensional approach of Selberherr *et al* [11].

Although the dielectric constant increases at high levels, the dielectric strength increase is a challenge. So, the added extra load results in a loss of dielectric strength. Efforts are required to gains specific information about QDs' diameters [12]. Therefore, it is necessary to understand the relation of QDs diameter-based bonds of nearest atoms. Controlling QD diameter-based materials links the QDs' potential to the QDs' diameter effect. Hence, the validity of our QD potential model [13] is tested. The energy gaps are utilized to calculate QD potential and predict new materials.

The verification of our model [13] is crucial to compute the diameter-based QD potential of $PbS_{0.25}Te_{0.75}$, $PbS_{0.5}Te_{0.5}$, and $PbS_{0.75}Te_{0.25}$ alloys for 68, 69, and 71 nm, respectively, employing the full-potential linearized augmented plane wave (FP-LAPW) method, in addition to investigating the optical dielectric constant and refractive index of optical properties utilizing empirical models.

9.2 Technique of chalcogenides-based quantum dots

The FP-LAPW method is employed to investigate Te concentration increase in optical studies of $PbS_{1-x}Te_x$ ternary alloys ($x = 0.0, 0.25, 0.5, 0.75, 1.0$). Zunger's special quasi-random structures (SQS) approach [14] is utilized to re-result the alloy's randomness of first shells around site. The size system is partitioned in a nanoscale optimization process by DFT. The physical properties of the alloys are not affected by periodicity beyond first shells, which are sufficient.

The FP-LAPW method in WIEN2K code [15] is utilized for the electronic and optical calculations. The generalized gradient approximation (GGA) [16], Engel and Vosko's generalized gradient approximation (EVGGA) formalism [17], and modified Becke Johnson (mBJ) [18] of exchange correlation potential are used for the total energy calculations. To exceed the underestimation utilizing LDA and GGA [19], the EVGGA and mBJ are employed, which is attributed to not correctly reproducing exchange correlation energy and its charge derivative. So, the EVGGA is ameliorated in mBJ, which is able to better reproduce the exchange potential and yield better splitting of bands. In FP-LAPW, the potential, charge density, and wave function are expanded by interstitial region of plane waves basis set in unit cell space and muffin-tin spheres of nonoverlapping spheres surrounding the atomic sites for spherical harmonic functions. The $l_{max} = 10$ value is a confined atomic sphere of wave functions. The muffin-tin radii are 1.95 for both Te and Pb, and 1.6 atomic units (a.u.) for S. The $K_{max} = 9.0/RMT$ of the cut-off for plane waves is chosen for wave function expansion into the interstitial region of PbS, PbTe, and $PbS_{1-x}Te_x$ alloy, but $G_{max} = 14\ (Ryd)^{1/2}$ is a Fourier expansion of charge density. The Brillouin zone irreducible wedge is attributed to 35 special k-points mesh of compounds and alloys with an exceptional, $x = 0.5$ to utilize 64 special k-points mesh. The 220 k-points mesh is utilized for binary and alloys with an exceptional $x = 0.5$ to utilize 216 k-points mesh for energy calculations. The converged self-consistent calculations are 10^{-5} Ry for total energy calculations.

9.3 Investigation of chalcogenides-based quantum dots

The four-fold coordination is attributed to covalent semiconductors because the low structure density and atomic nearest neighbors are bounded via hybridized and overlapped orbitals of tetrahedral shape, known as sp^3 hybrids. So, it is possible to tune QD diameter-based energy gaps. Table 9.1 displays calculated direct ($\Gamma \rightarrow \Gamma$) and indirect ($\Gamma \rightarrow L$) ($\Gamma \rightarrow X$) energy gaps employing EVGGA and mBJ for QD diameter-based $PbS_{1-x}Te_x$ alloys, along with experimental [20] and theoretical data [21, 22] for comparison. Our calculated values of the ($\Gamma \rightarrow \Gamma$) energy bandgap are slightly overestimated compared to other results. This could be attributed to using of the EVGGA and mBJ approximations. Therefore, $PbS_{1-x}Te_x$ alloys are regarded as direct energy gap semiconductors. The QD diameter-based energy gaps are necessary for applications in infrared light detection and generation. Table 9.1 presents a correlation of QDs diameter and energy gaps, as verified by figure 9.1.

The energy bandgaps between the valence band maximum at point Γ and the conduction band minimum at point X are computed based on the FP-LAPW. By means of our recent model [13], the QD potential has been evaluated according to the formula:

Table 9.1. Calculated QD diameter-based energy gaps of $PbS_{0.75}Te_{0.25}$, $PbS_{0.5}Te_{0.5}$, and $PbS_{0.25}Te_{0.75}$ alloys in eV compared to experimental and theoretical data.

QD diameter	E_g (Γ−Γ)	E_g (Γ−X)	E_g (Γ−L)
PbS	1.2448[C1] 1.1888[C2] 0.286[a] 0.989[b] 0.069[c]	7.1691[C1] 6.786[C2]	5.2538[C1] 4.6518[C2]
PbTe	1.2948[C1] 1.185[C2] 0.19[a] 0.5704[b] 0.032[c]	5.6827[C1] 5.1268[C2]	4.0139[C1] 3.8054[C2]
$PbS_{0.75}Te_{0.25}$	C_1 C_2	C_1 C_2	C_1 C_2
71	0.7895 0.9752	1.2505 1.4375	2.5299 2.678
73	0.9242 1.1681	1.3106 1.5433	2.53 2.7164
74	1.0012 1.2901	1.3355 1.6054	2.5004 2.7181
76	1.0461 1.3683	1.3414 1.6398	2.4573 2.6992
77	1.0682 1.42	1.3329 1.6576	2.4055 2.6697
78	1.0796 1.4509	1.3192 1.6615	2.3526 2.6317
$PbS_{0.5}Te_{0.5}$	C_1 C_2	C_1 C_2	C_1 C_2
69	0.8103 1.0532	1.2157 1.4425	2.6332 2.7619
71	0.9429 1.2492	1.2881 1.5677	2.6235 2.7972
73	1.0133 1.3622	1.3192 1.6367	2.5877 2.7937
74	1.0544 1.431	1.329 1.6734	2.5425 2.7717
75	1.0774 1.4737	1.326 1.6923	2.4935 2.7407
76	1.0891 1.4989	1.3156 1.6954	2.4435 2.703
$PbS_{0.25}Te_{0.75}$	C_1 C_2	C_1 C_2	C_1 C_2
68	1.0563 1.3687	1.4462 1.7495	2.9386 3.1281
69	1.1557 1.5321	1.4912 1.8491	2.9258 3.1614
71	1.2138 1.6326	1.5112 1.9053	2.892 3.1605
72	1.2447 1.6943	1.5121 1.9344	2.8463 3.1402
73	1.262 1.7321	1.506 1.9468	2.7941 3.1091
74	1.2611 1.7535	1.4868 1.948	2.7221 3.0677

[a] Reference [20] experiment.
[b] Reference [21] theory.
[c] Reference [22] theory.
[C1] calculated by EVGGA.
[C2] calculated by mBJ.

$$P_{QD} = \frac{b}{a}.\ E_{g\Gamma X}.\ 10^{-3}.\ \lambda \tag{9.1}$$

where constant refers to $\frac{b}{a}$ in eV^{-1} (see table 4 in reference [13]), energy gap refers to $E_{g\Gamma X}$ in eV, and parameter refers to λ for II–VI ($\lambda = 2$), III–V ($\lambda = 4$) and group-IV ($\lambda = 6$) materials in V.

There is a known correlation between the pressure effect and the QD's diameter. If the QD's diameter changes, the sp^3 covalent is affected. So, the QD potential difference is attributed to changes in the QD's diameter. Therefore, table 9.2 presents diameter-based QD potential. It is noted that different QD diameters employing Zunger's SQS approach reproduce the alloy's randomness [23].

The separation of the increasing and decreasing QD potential is critical to the QD's diameter. The correlation of transition pressure (P_t) and QD diameter is

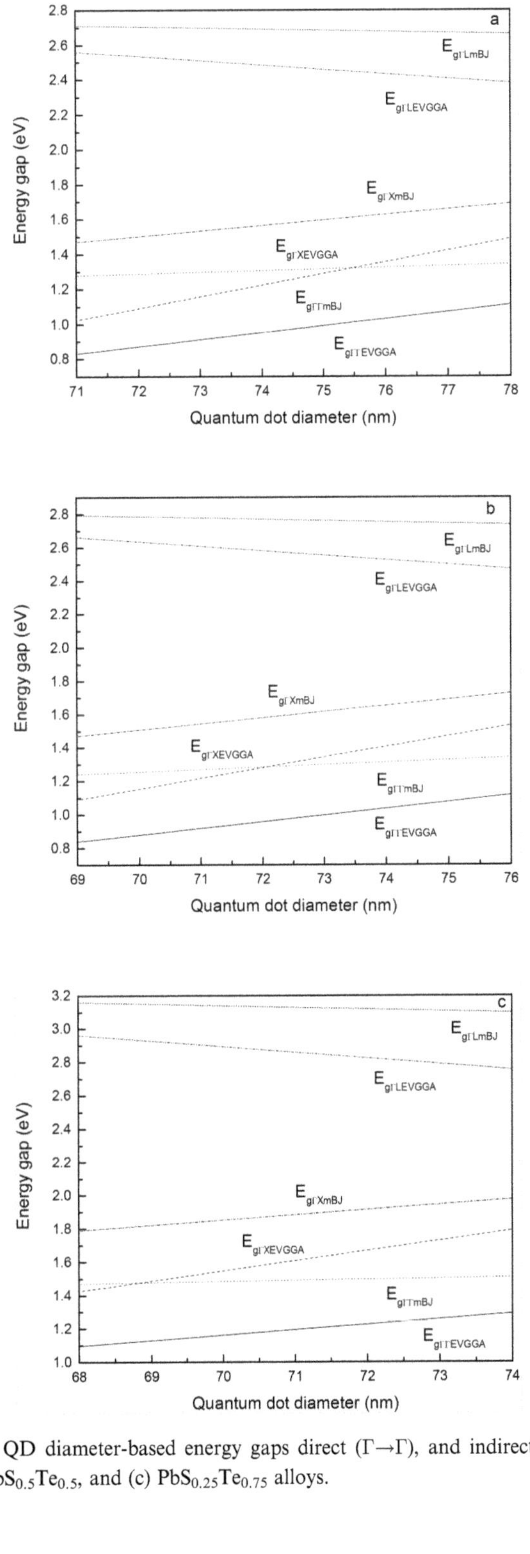

Figure 9.1. Calculated QD diameter-based energy gaps direct (Γ→Γ), and indirect (Γ→X) and (Γ→L) of (a) $PbS_{0.75}Te_{0.25}$, (b) $PbS_{0.5}Te_{0.5}$, and (c) $PbS_{0.25}Te_{0.75}$ alloys.

Table 9.2. Comparison of calculated QD diameter-based QDs of $PbS_{0.75}Te_{0.25}$, $PbS_{0.5}Te_{0.5}$, and $PbS_{0.25}Te_{0.75}$ alloys in mV in comparison to other data.

QD's diameter	P_{QD} cal.	P_{QD} [9]
$PbS_{0.75}Te_{0.25}$	C_1 C_2	1
71	0.450 0.555	1
73	0.526 0.658	1
74	0.570 0.735	
76	0.596 0.779	
77	0.608 0.809	
78	0.615 0.827	
$PbS_{0.5}Te_{0.5}$	C_1 C_2	
69	0.461 0.600	
71	0.537 0.712	
73	0.577 0.776	
74	0.601 0.815	
75	0.614 0.840	
76	0.620 0.854	
$PbS_{0.25}Te_{0.75}$	C_1 C_2	
68	0.602 0.780	
69	0.658 0.873	
71	0.691 0.930	
72	0.709 0.965	
73	0.718 0.987	
74	0.719 0.999	

C_1: calculated by EVGGA, C_2: calculated by mBJ.

studied in terms of molar free energies difference. The Gibbs free-energy difference (ΔG_t) has tetrahedral coordination at the QD diameter, which is given by $\Delta G = \Delta H - T\Delta S$ (in kJ.mol^{-1}), where enthalpy refers to H, temperature refers to T, and entropy refers to S. An inverse correlation is observed between energies and bond length. Changing the QD diameter-based QD potential is verified by the energy gap change at points Γ−Γ, Γ−X, and Γ−L, as illustrated in table 9.1. The correlation is directly (table 9.2) verified by figure 9.2 and is linear for $PbS_{1-x}Te_x$ alloys. As a result, a QD potential fluctuation appears. Our QD potential is in accord with other work [9]. Therefore, the QD potential variation indicates that electrons are tunneling across the QDs.

One of the important optical properties is refractive index n. Theoretically, there are two approaches: first, the n is related to local polarizability and density, and second the n is related to band structure within dielectric constant [24]. Therefore, efforts have been made to relate n and E_g within specific relationships [25]. These relations of n are independent of incident photon energy and temperature. So, different relationships of n and E_g are stated. Ravindra *et al* [26] have investigated n in terms of E_g:

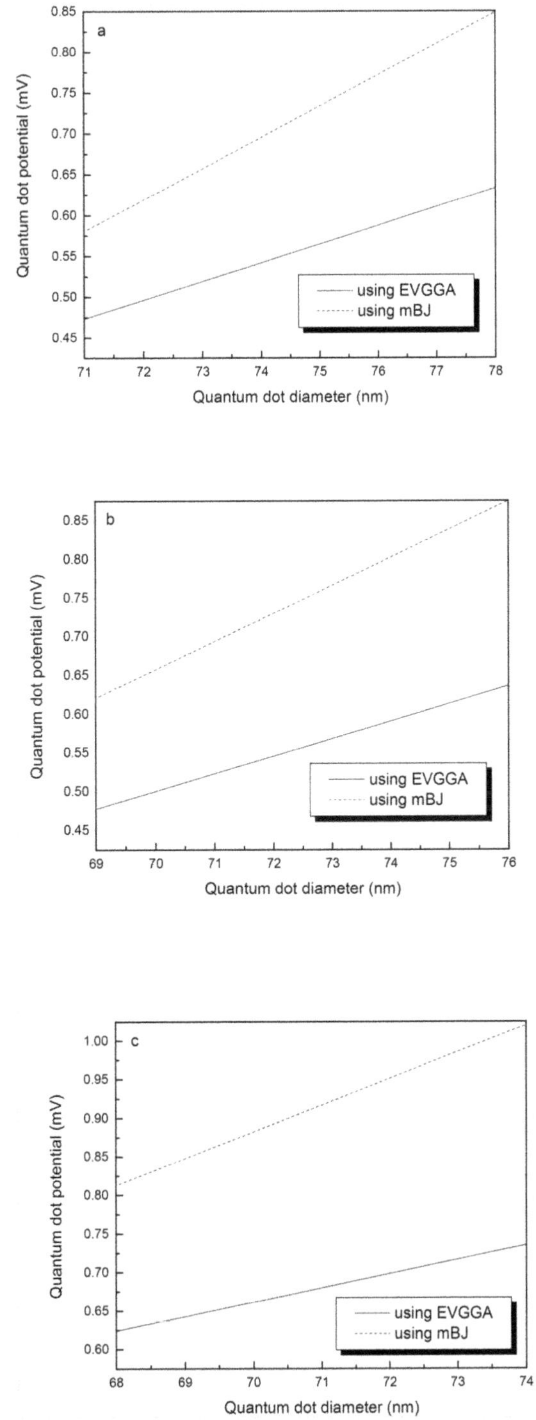

Figure 9.2. Diameter-based QDs of (a) $PbS_{0.75}Te_{0.25}$, (b) $PbS_{0.5}Te_{0.5}$, and (c) $PbS_{0.25}Te_{0.75}$ alloys.

$$n = \alpha + \beta E_g \tag{9.2}$$

where $\alpha = 4.048$ and $\beta = -0.62\ \text{eV}^{-1}$. Herve and Vandamme [27] have proposed an empirical relation, as follows:

$$n = \sqrt{1 + \left(\frac{A}{E_g + B}\right)^2} \tag{9.3}$$

where $A = 13.6$ eV and $B = 3.4$ eV. For group II–IV semiconductors, Ghosh *et al* [28] have published an empirical relationship based on the band structure and quantum dielectric considerations of Penn [29] and Van Vechten [30]:

$$n^2 - 1 = \frac{A}{(E_g + B)^2} \tag{9.4}$$

where $A = 8.2E_g + 134$, $B = 0.225E_g + 2.25$, and $(E_g + B)$ refer to E_g. So, the three empirical models of n are utilized. Figure 9.3 illustrates the accepted results, and table 9.3 presents the calculated optical dielectric constants and refractive indices.

The optical dielectric constant ε_∞ depends on n. Note that $\varepsilon_\infty = n^2$ [33]. It is obvious that n utilizing Ghosh *et al's* [28] model and mBJ approximation is suitable for PVs. Once again, the QD diameter-based $PbS_{1-x}Te_x$ alloys are noticed and there are small QD diameter-based refractive index shifts to blue–green that increase the PVs' efficiencies by increasing absorption and decreasing reflection.

Finally, the FP-LAPW method has proved to be an excellent way to investigate the electronic and optical properties of $PbS_{1-x}Te_x$ alloys and present that 74, 76, and 78 nm for $PbS_{0.25}Te_{0.75}$, $PbS_{0.5}Te_{0.5}$, and $PbS_{0.75}Te_{0.25}$ alloys, respectively, utilizing Ghosh *et al's* model are more suitable for PV applications to explore new QDs trends and alloys. In addition, mBJ is more suitable for QD potential.

Problems

1. What are chalcogenides-based QDs? And, how are they synthesized?
2. How do the optical properties of chalcogenides-based QDs change as a function of their size and composition?
3. What are the challenges associated with investigating the optical properties of chalcogenides-based QDs using first-principles calculations?
4. What are some potential applications of chalcogenides-based QDs in optoelectronics? And, how can first-principles calculations be used to optimize their performance?
5. How can experimental measurements of the optical properties of chalcogenides-based QDs be used to validate and improve the accuracy of first-principles calculations?

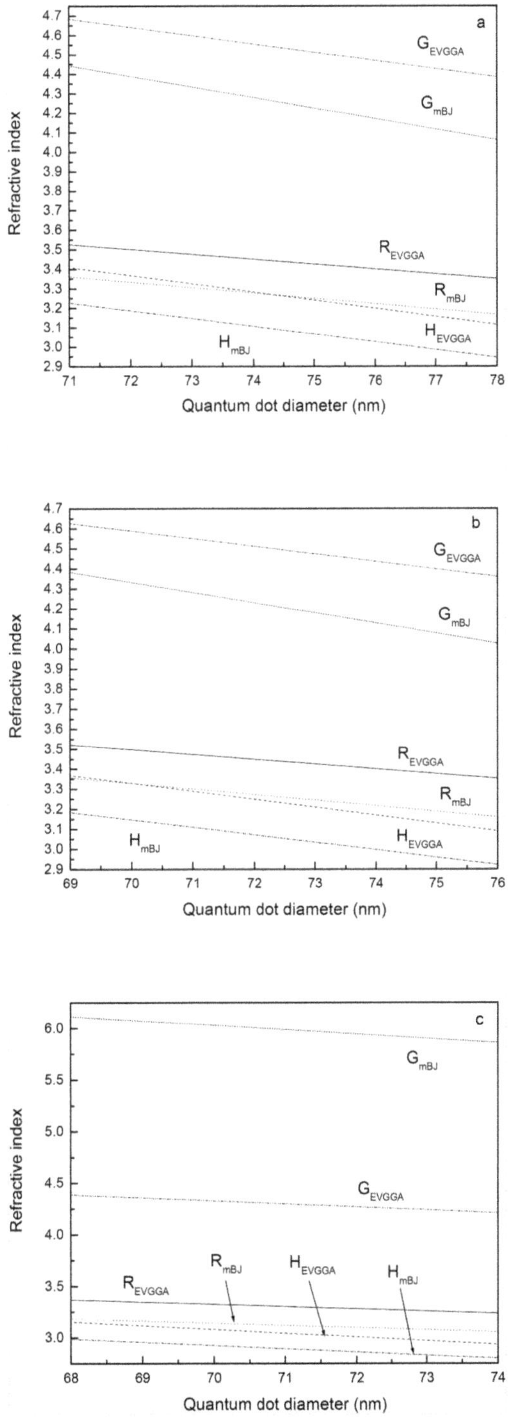

Figure 9.3. QD diameter-based refractive index (*n*) of (a) $PbS_{0.75}Te_{0.25}$, (b) $PbS_{0.5}Te_{0.5}$, and (c) $PbS_{0.25}Te_{0.75}$ alloys. R: Ravindra *et al* [26], H: Herve and Vandamme [27], and G: Ghosh *et al* [28].

Table 9.3. Calculated QD diameter-based refractive indices of $PbS_{0.75}Te_{0.25}$, $PbS_{0.5}Te_{0.5}$ and $PbS_{0.25}Te_{0.75}$ alloys utilizing Ravindra *et al's* [26], Herve and Vandamme's [27] and Ghosh *et al's* [28] models, corresponding to optical dielectric constant.

QD's diameter	n	ε_∞
PbS	3.27[a,1] 3.31[a,2] 3.09[b,1] 3.12[b,2] 4.26[c,1] 4.30[c,2]	10.69[a,1] 10.95[a,2] 9.54[b,1] 9.73[b,2] 18.14[c,1] 17.30[c,2]
PbTe	4.16[d]	10.49[a,1] 10.95[a,2] 9.36[b,1] 9.79[b,2] 38.68[c,1] 39.81[c,2]
$PbS_{0.75}Te_{0.25}$	3.24[a,1] 3.31[a,2] 3.06[b,1] 3.13[b,2] 6.22[c,1] 6.31[c,2]	12.60[a,1] 11.83[a,2] 11.49[b,1] 10.62[b,2] 21.90[c,1] 20.16[c,2]1
71	5.98[e]	12.04[a,1] 11.02[a,2] 10.89[b,1] 9.85[b,2] 20.61[c,1] 18.66[c,2]
73	3.55[a,1] 3.44[a,2] 3.39[b,1] 3.26[b,2] 4.68[c,1] 4.49[c,2]	11.69[a,1] 10.49[a,2] 10.49[b,1] 9.36[b,2] 21.71[c,1] 17.80[c,2]
74	3.47[a,1] 3.32[a,2] 3.30[b,1] 3.14[b,2] 4.54[c,1] 4.32[c,2]	11.49[a,1] 10.17[a,2] 10.30[b,1] 9.12[b,2] 19.62[c,1] 17.30[c,2]
76	3.42[a,1] 3.24[a,2] 3.24[b,1] 3.06[b,2] 4.66[c,1] 4.22[c,2]	11.42[a,1] 9.98[a,2] 10.24[b,1] 8.94[b,2] 19.44[c,1] 16.97[c,2]
77	3.39[a,1] 3.19[a,2] 3.21[b,1] 3.02[b,2] 4.43[c,1] 4.16[c,2]	11.35[a,1] 9.85[a,2] 10.17[b,1] 8.82[b,2] 19.36[c,1] 16.81[c,2]
78	3.38[a,1] 3.16[a,2] 3.20[b,1] 2.99[b,2] 4.41[c,1] 4.12[c,2]	12.53[a,1] 11.49[a,2] 11.42[b,1] 10.30[b,2] 21.71[c,1] 19.53[c,2]
$PbS_{0.5}Te_{0.5}$	3.37[a,1] 3.14[a,2] 3.19[b,1] 2.97[b,2] 4.40[c,1] 4.10[c,2]	11.97[a,1] 10.69[a,2] 10.75[b,1] 9.54[b,2] 20.43[c,1] 18.06[c,2]
69	3.54[a,1] 3.39[a,2] 3.38[b,1] 3.21[b,2] 4.66[c,1] 4.42[c,2]	11.62[a,1] 10.24[a,2] 10.43[b,1] 9.12[b,2] 19,89[c,1] 17.30[c,2]
71	3.46[a,1] 3.27[a,2] 3.28[b,1] 3.09[b,2] 4.52[c,1] 4.25[c,2]	11.49[a,1] 9.98[a,2] 10.30[b,1] 8.88[b,2] 19.53[c,1] 16.89[c,2]
73	3.41[a,1] 3.20[a,2] 3.23[b,1] 3.02[b,2] 4.46[c,1] 4.16[c,2]	11.42[a,1] 9.79[a,2] 10.17[b,1] 8.76[b,2] 19.36[c,1] 16.64[c,2]
74	3.39[a,1] 3.16[a,2] 3.21[b,1] 2.98[b,2] 4.42[c,1] 4.11[c,2]	11.35[a,1] 9.67[a,2] 10.11[b,1] 8.70[b,2] 19.27[c,1] 16.48[c,2]
75	3.38[a,1] 3.13[a,2] 3.19[b,1] 2.96[b,2] 4.40[c,1] 4.08[c,2]	11.49[a,1] 10.17[a,2] 10.30[b,1] 9.12[b,2] 19.53[c,1] 37.94[c,2]
76	3.37[a,1] 3.11[a,2] 3.18[b,1] 2.95[b,2] 4.39[c,1] 4.06[c,2]	11.08[a,1] 9.54[a,2] 9.85[b,1] 8.58[b,2] 18.74[c,1] 36.48[c,2]
$PbS_{0.25}Te_{0.75}$	3.39[a,1] 3.19[a,2] 3.21[b,1] 3.02[b,2] 4.42[c,1] 6.16[c,2]	10.82[a,1] 9.18[a,2] 9.67[b,1] 8.29[b,2] 18.31[c,1] 35.64[c,2]
68	3.33[a,1] 3.09[a,2] 3.14[b,1] 2.93[b,2] 4.33[c,1] 6.04[c,1]	10.69[a,1] 8.94[a,2] 9.54[b,1] 8.12[b,2] 18.14[c,1] 35.16[c,2]
69	3.29[a,1] 3.03[a,2] 3.11[b,1] 2.88[b,2] 4.28[c,1] 5.97[c,2]	10.62[a,1] 8.82[a,2] 9.48[b,1] 8.00[b,2] 17.97[c,1] 34.81[c,2]
71	3.27[a,1] 2.99[a,2] 3.09[b,1] 2.85[b,2] 4.26[c,1] 5.93[c,2]	10.56[a,1] 8.76[a,2] 9.42[b,1] 7.95[b,2] 17.89[c,1] 34.69[c,2]
72	3.26[a,1] 2.97[a,2] 3.08[b,1] 2.83[b,2] 4.24[c,1] 5.90[c,2]	
73	3.25[a,1] 2.96[a,2] 3.07[b,1] 2.82[b,2] 4.23[c,1] 5.89[c,2]	
74		

[a] Reference [26].
[b] Reference [27].
[c] Reference [28].
[d] Reference [31] experiment.
[e] Reference [32] experiment.
[1] Calculated by EVGGA.
[2] Calculated by mBJ.

References

[1] Agrawal G P and Dutta N K 1993 *Semiconductor Lasers* (New York: Van Nostrand Reinhold)
[2] Nair P K, Ocampo M, Fernandez A and Nair M T S 1990 *Sol. Energy Mater.* **20** 235
[3] Zhang S B and Cohen M L 1987 *Phys. Rev.* B **35** 7604
[4] Green M A 2003 *Third Generation Photovoltaics: Ultra-High Efficiency at Low Cost* (Berlin: Springer)
[5] Nelson J 2003 *The Physics of Solar Cells* (London: Imperial College Press)
[6] Luo J, Bai H, Yang P and Cai J 2015 *Mater. Sci. Semicond. Process.* **34** 1
[7] El-Rabaie S, Taha T A and Higazy A A 2015 *Mater. Sci. Semicond. Process.* **30** 631
[8] Strong W H, Forbes D V and Hubbard S M 2014 *Mater. Sci. Semicond. Process.* **25** 76
[9] Udipi S, Vasileska D and Ferry D K 1996 *Superlattices Microstruct.* **20** 343
[10] Scharfetter D L and Gummel H K 1969 *IEEE Trans. Electron Devices* **ED-16** 64
[11] Selberherr S, Shutz A and Potzl H W 1980 *IEEE Trans. Electron Devices* **ED-27** 1540
[12] Al-Douri Y, Baaziz H, Charifi Z, Khenata R, Hashim U and Al-Jassim M 2012 *Renewable Energy* **45** 232
[13] Al-Douri Y 2009 Quantum dot modeling of semiconductors in: A H Reshak ed *Advances in Condensed Matter Physics* (Kerala, India: Research Signpost)
[14] Zunger A, Wei S-H, Ferreira L G and Bernard E 1990 *Phys. Rev. Lett.* **65** 353
[15] Blaha P, Schwarz K, Madsen G K H, Kvasnicka D and Luitz J 2001 *WIEN2K* (Wien, Austria: Technische Universitat)
[16] Perdew J P, Burke S and Ernzerhof M 1996 *Phys. Rev. Lett.* **77** 3865
[17] Engel E and Vosko S H 1993 *Phys. Rev.* B **47** 13164
[18] Becke A D and Johnson E R 2006 *J. Chem. Phys.* **124** 221101
[19] Dufek P, Blaha P and Schwarz K 1994 *Phys. Rev.* B **50** 7279
[20] Strehlow W H and Cook E L 1973 *J. Phys. Chem. Ref. Data* **2** 163
[21] Rached D, Rabah M, Benkhettou N, Driz M and Soudini B 2003 *Physica* B **337** 394
[22] Delin A, Ravindran P, Eriksson O and Wills J M 1998 *Int. J. Quantum Chem.* **69** 349
[23] Zunger A, Wei S-H, Ferreira L G and Bernard E 1990 *Phys. Rev. Lett.* **65** 353
[24] Balzaretti N M and da Jornada J A H 1996 *Solid State Commun.* **99** 943
[25] Al-Douri Y, Feng Y P and Huan A C H 2008 *Solid State Commun.* **148** 521
[26] Ravindra N M, Auluck S and Srivastava V K 1979 *Phys. Status Solidi* B **93** K155
[27] Herve P J L and Vandamme L K J 1995 *J. Appl. Phys.* **77** 5476
[28] Ghosh D K, Samanta L K and Bhar G C 1984 *Infrared Phys.* **24** 34
[29] Penn D R 1962 *Phys. Rev.* **128** 2093
[30] Van Vechten J A 1969 *Phys. Rev.* **182** 891
[31] Zemel J N, Jensen J D and Schoolar R B 1965 *Phys. Rev.* **140** A330
[32] Weiting F and Yixun Y 1990 *Infrared Phys.* **30** 371
[33] Samara G A 1983 *Phys. Rev.* B **27** 3494

Chapter 10

Artificial intelligence-based quantum dots

This chapter explores the application of artificial intelligence (AI) techniques in the design, synthesis, and optimization of quantum dots (QDs). The key aspects of this study include AI-driven design, which utilizes machine-learning algorithms to predict the properties of QDs based on their size, composition, and synthesis conditions. AI can identify optimal conditions for achieving desired optical and electronic characteristics. Meanwhile, synthesis optimization is used to optimize the synthesis process of QDs. AI models predict optical properties, which are based on particle size, shape and material composition, enabling rapid screening of potential quantum dot configurations.

10.1 Introduction

Studying QDs using AI is a rapidly growing field, which has shown a great promise in the design and discovery of novel materials. By combining AI with quantum mechanical simulations, it is possible to accelerate the discovery of new QDs with desirable properties for various applications. The process of designing QDs using AI typically involves training machine-learning algorithms on a large dataset of known QDs with known properties. These algorithms use this information to predict the properties of new QDs that have not yet been synthesized or characterized experimentally.

A recent study [1] has demonstrated the use of carbon quantum dots (CQDs) to predict the optical properties for sensing materials. The authors have explored wide utilizing of sensing platform. They described attractive advancements of CQDs and optical investigations with a focus on preparation technique, luminescence-dependence of chemical composition, and size, including their applications. They also discussed the future challenges in developing sensors. By reducing the need for expensive and time-consuming experimental synthesis, analysis, and characterization, AI can help to streamline the development of QDs-based devices for optoelectronic and sensing applications.

doi:10.1088/978-0-7503-5704-3ch10

There are several challenges are associated with using AI in QD design. The main challenge is the need for large, high-quality datasets of QDs with specific properties, which cannot be easily obtained experimentally. Another challenge is the AI models' interpretability given that it can be difficult to understand how the models make their investigations and predictions, and whether they are reliable.

Overall, AI-based QD design represents a promising and rapidly developing area of research, with a great potential to accelerate the discovery of new materials with desirable properties for a wide range of applications. As the field evolves, it will be necessary to address the challenges associated with AI-based approaches for ensuring that the resulting QDs are not only optimized for specific applications but exhibit desirable biological and environmental safety profiles.

10.2 Principles of artificial intelligence-based quantum dots

The principles of AI-based QDs are related to using machine-learning algorithms and quantum mechanical simulations to investigate and design the properties of QDs [2]. This involves the following:

1. Data collection: Large datasets of QDs with specific properties such as composition, size, and physical properties are required.
2. Data preprocessing: The data must be processed, normalized, and cleaned to remove any errors or outliers.
3. Feature extraction: Datasets such as energy bandgap, exciton binding energy, and energy levels need to be extracted to describe the QDs' properties.
4. Model training: The machine-learning algorithms are trained on datasets utilizing unsupervised or supervised machine learning to identify the relationships and patterns between the QDs' properties and features.
5. Model evaluation: The effectiveness of a model is evaluated utilizing cross-validation methods such as *k*-fold validation to assess its generalizability and accuracy.
6. Prediction: Once the model is validated and trained, it can predict the new QDs properties that have not been prepared, analyzed, or characterized experimentally.

By influencing the AI power and quantum mechanical simulations, we can accelerate the design and discover novel QDs with optical and electronic properties, which can be utilized in different quantum computing, sensing, and optoelectronic applications. The principles of AI-based QDs rely on the availability of large high-quality datasets in addition to the ability to simulate and investigate a system's quantum mechanical behavior. To continue the development, it is necessary to address the challenges to ensure that the resulting QDs are not only optimized for specific applications but exhibit desirable biological and environmental profiles.

10.3 Fundamental concepts of artificial intelligence-based quantum dots

AI-based QDs are designed using fundamental concepts from both AI and quantum mechanics [3]. Some of the key fundamental concepts include:

1. Machine-learning algorithms: These are used to analyze large datasets of QDs with specific properties to learn the relationships between different properties and implicit features. This allows us to predict specific properties of new QDs that have not been prepared, analyzed, or characterized experimentally.
2. Quantum mechanical simulations: These are used to calculate the optical and electronic properties of QDs based on their structure, composition, and size. This investigation is critical to predict the properties of novel QDs, such as the system's behavior at the quantum level.
3. Feature extraction: This identifies the features and properties of QDs relevant to predicting their properties, including features such as the electronic structure, composition, shape, and size of QDs.
4. Model training: The machine-learning models are trained on a QD dataset using unsupervised and supervised learning methods, identifying the parameters of the optimal model to minimize error between the actual and predicted properties of the QD dataset.
5. Prediction: When the machine-learning model has been trained, it can predict new QD properties based on their composition, size, and so on.

By gathering the fundamental concepts, AI-based QDs offer a powerful approach to optimize and design novel QDs for quantum computing, sensing, and optoelectronic applications. The significant challenge is to get efficient quantum mechanical simulations and high-quality QD datasets.

10.4 Characteristics of artificial intelligence-based quantum dots

AI-based QDs have several unique characteristics that distinguish them from traditional QDs [4]. Some of the key characteristics include:

1. Predictive power: One of benefits of AI-based QDs is that they can predict the properties of QDs with high accuracy. To analyze large datasets of QDs, the machine-learning algorithms identify relationships and patterns to predict QD properties that have not been prepared, analyzed, or characterized experimentally.
2. Optimization potential: AI-based QDs offer a potential for optimization of QD properties basing on specific applications. Machine learning can be used to predict the QD properties of specific structures, compositions, and sizes to optimize the performance of QDs for quantum computing, optoelectronic, and sensing applications.
3. Speed and efficiency: The use of machine learning and quantum mechanical simulations allows for rapid checking of QDs, significantly reducing the time

and resources necessary for experimental preparation, analysis, and characterization.

4. Customizability: AI-based QDs offer a high degree of customizability, permitting QDs to be designed specific properties and applications, including structure, composition, and size tuning of QDs to optimize their optical and electronic properties.
5. Integration potential: The predictive customizability and power of AI-based QDs make them an attractive choice for integration into a range of applications, such as quantum computers, LEDs, and solar cells.

In general, AI-based QDs present a powerful approach for the optimization and design of QDs for various applications. However, more research is needed to develop efficient and accurate quantum mechanical simulations to obtain high-quality datasets of QDs to realize their potential.

10.5 Fabrication of artificial intelligence-based quantum dots

The fabrication of AI-based QDs involves two main steps: data generation and the use of machine-learning algorithms to predict the properties of novel QDs [5]:

1. Data generation: To generate data for AI-based QDs, researchers use a combination of experimental measurements and quantum mechanical simulations. Experimental measurements such as emission spectroscopy and absorption can characterize the properties of QDs. Meanwhile, quantum mechanical simulations such as density functional theory can predict the optical and electronic properties of QDs with specific structures, compositions, and sizes.
2. Machine learning: Once a dataset of QDs has been generated, machine-learning algorithms can identify relationship between the QDs' properties and their features.

The fabrication of AI-based QDs can also involve the use of advanced synthesis techniques to tailor the size, composition, and structure of the QDs to optimize their properties for specific applications. This can include the use of solution-phase synthesis, molecular beam epitaxy, and colloidal synthesis techniques.

The manufacturing of AI-based QDs requires a combination of experimental techniques and advanced computational techniques, in addition to efficient machine-learning algorithms and accurate development to predict the properties of QDs.

10.6 Preparation of artificial intelligence-based quantum dots

The preparation of AI-based QDs includes the selection of materials, QDs synthesis, characterization, and the use of machine-learning algorithms for predicting the QD's properties [6]:

1. Material selection: This selects the materials for QDs based on their properties, such as cadmium selenide (CdSe), lead sulfide (PbS), or indium

phosphide (InP), which are used due to their tunable optical and electronic properties.
2. Synthesis: QDs are prepared by molecular beam epitaxy and colloidal synthesis, which allow for precise control over the composition, shape, and size of QDs.
3. Characterization: The QDs are characterized by transmission electron microscopy, x-ray diffraction, and absorption and emission spectroscopy to provide information about the size, shape, crystal structure, and the optical properties of the QDs.
4. Machine learning: Once a dataset of QDs has been generated, machine-learning algorithms can be utilized to predict the properties of QDs. These algorithms utilize statistical methods to identify the relationships between QDs' properties and their features.
5. Optimization: The properties of QDs are optimized by adjusting their composition, shape, and size based on the prediction of machine-learning algorithms. This process is used to design QDs for various quantum computing, LED, and solar cells applications.

The preparation of AI-based QDs includes a combination of machine learning, characterization, synthesis, and materials selection to design QDs for different applications.

10.7 Fabrication methods for artificial intelligence-based quantum dots

There are several fabrication methods to synthesize AI-based QDs [7], including:

1. Colloidal synthesis: This uses a solvent, a precursor material, and a stabilizing agent. The precursor material is added to the solvent and a stabilizing agent is added to prevent the particles from aggregating. The particles are heated to promote the nucleation and growth of QDs.
2. Molecular beam epitaxy: A substrate is first placed in a high-vacuum chamber. The substrate is then bombarded with atoms or molecules of desired materials that self-assemble into QDs on the substrate.
3. Sol-gel method: This uses a precursor material dissolved in a solvent. The mixture is heated to promote hydrolysis and condensation reactions, which result in the formation of a gel-like material that is dried and calcined to form QDs.
4. Laser ablation: This uses a high-powered laser to vaporize a target material. The vaporized material then condenses into QDs on a substrate.
5. Electrochemical synthesis: This uses an electrode and a precursor material in a solution. When a voltage is applied to the electrode, it promotes the reduction of the precursor material that forms QDs.

The choice of fabrication method for AI-based QDs depends on several factors, such as the desired composition, shape, and size of QDs, in addition to the QD's applications.

10.8 Optical properties and investigations of artificial intelligence-based quantum dots

The optical properties of AI-based QDs can be tuned based on their composition, shape, and size, making them highly attractive for a wide range of applications, including energy conversion, imaging, and sensing [8]. One of the most notable optical properties of AI-QDs is their size-dependent energy bandgap. Due to quantum confinement effects, AI-QDs exhibit discrete energy levels that can be tuned by varying the size of the QDs. This size-dependent energy bandgap can be probed using optical spectroscopy techniques, such as absorption, photoluminescence, and Raman spectroscopy.

Absorption spectroscopy can be utilized to investigate energy levels in AI-QDs by measuring the light absorption at different wavelengths. The absorption spectrum provides information about the QDs' composition and size, in addition to electronic structure and energy levels.

Photoluminescence spectroscopy measures the light emission of material after it has been excited by a light. In AI-QDs, photoluminescence is caused by recombination of electron-hole pairs that leads to light emission at a specific wavelength. The emission spectrum provides information about QDs composition and size, in addition to their optical and electronic properties.

Raman spectroscopy is used to probe the optical properties of AI-QDs, which involves the measurement of scattered light from a sample after it has been excited by a laser beam. The scattered light provides information about the QDs lattice structure and vibrational modes, which identify their crystalline structure and composition.

Finally, the AI-QDs optical properties make them highly attractive for a wide range of applications, including energy conversion, imaging, and sensing, which are attributed to controlling their optical properties and optimizing their performance.

Problems

1. How does AI aid in the optimization of QD synthesis and properties?
2. What are some potential applications of AI-based QDs in fields such as biomedicine or renewable energy?
3. How do the optical properties of AI-based QDs compare to traditional QDs synthesized through other methods?
4. Can AI be used to predict the behavior of QDs under various environmental conditions, such as temperature or pH?
5. What challenges still exist in the fabrication and characterization of AI-based QDs? And, how are researchers working to overcome them?

References

[1] Cruz A A C, Carneiro S V, Pontes S M A, Oliveira J J P, Lima J P O, Costa V M, Fechine L M U D, Clemente C S, Freire R M and Fechine P B A 2023 *Encycl. Sens. Biosens.* **2** 542

[2] Ma Y-F, Wang Y-M, Wen J, Li A, Li X-L, Leng M, Zhao Y-B and Lu Z-H 2023 *J. Electron. Sci. Technol.* **21** 100189
[3] Li Z, Zhang S and Mohammed B O 2023 *Mater. Sci. Eng.* B **294** 116526
[4] Kour S *et al* 2023 *J. Drug Delivery Sci. Technol.* **83** 104392
[5] Han D, Chen Y, Li D, Dong H, Xu B, He X and Sang S 2023 *Sens. Actuators* B **379** 133197
[6] Sohail A and Ashiq U 2023 *Sens. Int.* **4** 100212
[7] Xie Y, Sattari K, Zhang C and Lin J 2023 *Prog. Mater. Sci.* **132** 101043
[8] Costanzo H, Gooch J and Frascione N 2023 *Talanta* **253** 123945

www.ingramcontent.com/pod-product-compliance
Ingram Content Group UK Ltd.
Pitfield, Milton Keynes, MK11 3LW, UK
UKHW051849100926
8556IPUK00005B/531

9 780750 357029